THE UFO LIE

70 years of psychological warfare

EDITH VAN HERWI

INTRODUCTION

November 5, 1990 at 19h in Lagny sur Marne, while we go out, my mother, my brother and I a taxi in front of our house of the time, we notice the presence of a huge triangular gear a few hundred meters above us.
The UFO remained there for a few minutes, a kind of purring emanating from him, then climbed to cross the sound barrier a few kilometers away.
From that day, I started to believe, and my mother, we were not alone in the universe.
I even thought that I had been contacted by the entities that came to visit us.
Thirty years later, the archival documents, testimonies, declassification and the deductions made to understand that the aliens do not exist to me and millions of people had made and are still being manipulated to just continue 'believe in'.
Believing that we are surrounded in an open system possible planets to conquer while my research published on my blog " Cyprustar " prove that this is not the case.
The subject of the Earth paradigm will not be addressed in this disclosure, the truth about UFO being already sufficient to be treated here.
My mother, who died in 1998, did not have time to know the truth.
This book is dedicated to him and to all victims of the UFO phenomenon.

1 / UFO IN THE ART AND HISTORY?

We will see in the examples that follow, that no UFO appears in paintings, tapestries, engravings in history and art.

1 / Tapestry Beaune – France

Two tapestries of the 14th century the basilica of Notre-Dame in Beaune, depict the life of Mary. One can clearly see a kind of saucer. However according to the Beaune tradition, it is the representation of the hat of Cardinal Jean Rolin, Sponsor, ousted during the annexation of Burgundy by Louis XI.

2 / The crucifixion – Kosovo

"Crucifixion" dates from 1350 and is located in Kosovo Visoki Decani monastery, placed above the altar.
This also follows the crucifixion common iconographic model of the Middle Ages. The "Cross Deposition" of Benedetto Antelami in Parma dome in particular, like the crucifixion Visoki Decani from below:

On the edges of the composition, in the same position as in the fresco of the Visoki Decani, the Sun and the Moon are represented as human witnesses of the crucifixion, as in the previous table.
In both works of art, the characters that represent the sun and the moon turn to the cross in the center of the composition.

3 / Saint Announcement Emidius

Carlo Crivelli painting (1430-1495) "the Holy Announcement Emidius" (1486) shown at the National

Gallery, London. A bright disk top of the table, by magnifying one can see angels swirling inside this big cloud in a circle …

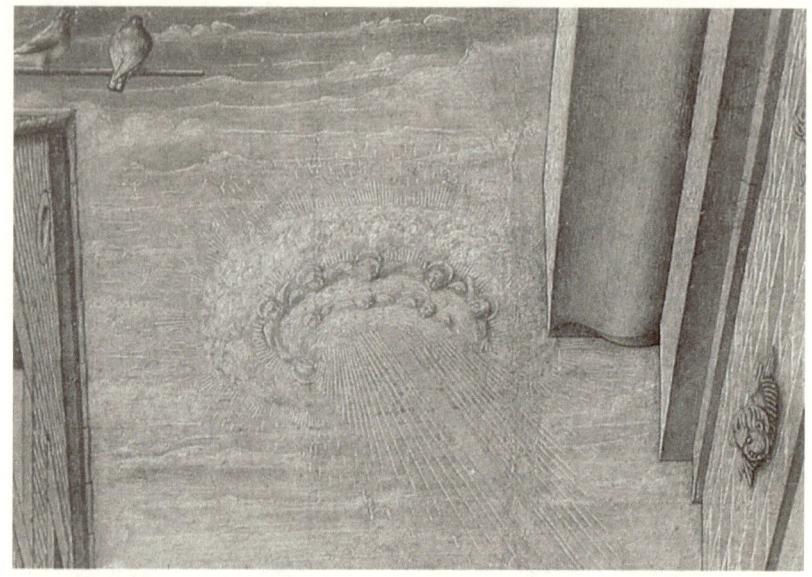

4 / The Miracle of the Snow – Italy

The scene depicts Jesus and Mary over what appear to

be lenticular clouds. The painting "Miracle of the Snow" and was painted by Masolino Da Panicale (1383-1440) and is visible in the church of Santa Maria Maggiore, Florence, Italy. There is no UFO in the table.
The legend of summer snowfall on the Esquiline is a story written a thousand years after the era where the miracle took place, and the clouds that have caused so much discussion that some ufologists fall in the style of different artists of the first half of the fifteenth century.

5 / The Madonna and St. Giovannino - Italy

Foto: © Diego Cuoghi, 2003

The "Madonna and St. Giovannino" was painted in the 15th century by Domenico Ghirlandaio (1449-1494) and is part of the collection "Loeser" (Palazzo Vecchio).

Above the shoulder of Mary, it seems that a man looks at what he seems to be for the UFO believers a flying disc.

In this painting attributed to Sebastiano Mainardi, an artist from the circle of Ghirlandaio, there is no UFO.

Foto: Diego Cuoghi 2003

The three little stars in this picture, accompany most of the Nativity, were often used to symbolize the triple virginity of Mary; the shepherd who looks at the appearance in the sky protecting his eyes from the hand is similar to many other extracts of paintings of the same subject;
the luminous cloud comes from the story of the Nativity in the apocryphal Protovangelo di Giacomo.
The same Protoevangelium, which contains the stories of the childhood of the Madonna is one of the most quoted texts in the definition of the dogma of Mary's virginity.

6 / Mary at the tomb of Jesus

Detail of the cover of a reliquary (Museo Sacro Vaticano, Rome)

This image shows Mary at the tomb of Jesus ... The object under which is the tomb of Jesus is the Holy Sepulcher dome. Not a UFO.

7 / Tokens "OPPORTVNVS ADEST"

The scene comes from Roman mythology. The object falling from the sky is not an UFO but the shield of Jupiter (Ancile). It was a gift from the Roman king Numa Pompilius.
The promise of God was supported by the sacred shield. The phrase "OPPORTUNUS ADEST" means:
"He is present in our time of need."

It's a call for protection.

8 / The Baptism of Christ

"The Baptism of Christ" by artist flemish Aert de Gelder painted in 1710 and visible in the Fitzwilliam Museum, Cambridge. A disc of light rays above the Baptist Jesus. The framework represents the "Baptism of Christ"; so we can compare it with many others on the same subject. We see immediately that one of the key elements of the composition is the intervention of God in the Baptism of the scene:

"And immediately, up from the water, he saw the heavens torn apart and the Spirit like a dove descending on him, and a voice came from heaven:" You are my beloved Son, you have all my support. " "(Mark 1.10). "As soon as Jesus was baptized, he went out of the water. And lo, the heavens were opened, and he saw the Spirit of God descending like a dove and lighting on him. And a

voice came from heaven these words: this is my beloved Son, in whom I am well pleased. " (Matthew 3,16-17).
"It happened, when all the people had been baptized and when Jesus was baptized too, was praying, the sky opened, and the Holy Spirit descended upon him in bodily form like a dove. And a voice came from heaven: "You are my son; I, today I have begotten you." "(Luke 3:21).

"And John bare record, saying:" I saw the Spirit descending like a dove from heaven and stay on him. "(John 1.32)

The Holy Spirit appears in the form of a dove, often within a circle of light which leave the rays that symbolize the divine Grace. Here we see the classic representation of the third person of the Trinity. In this case, it's "Disputation of the Sacrament" Raffaello and "Baptism of Christ" the Perugino:

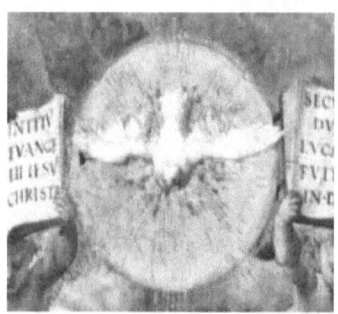 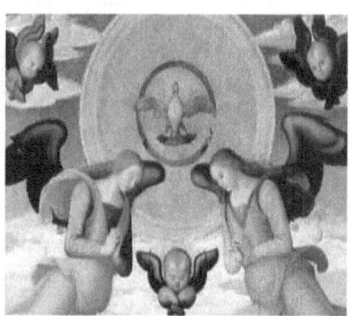

It is not hard to imagine why in so many cultures and religions, the deity is represented within a luminous disc. The early gods were the sun and the moon, appearing as a bright disk respectively dominate the day and night.
An example may be the god Aton imposed by Amenhetep IV (Pharaoh "heretic" who took the name of Akhnaton), which was represented in the form of solar disk which came out of the rays.

9 / Book of Morality – France

Image from the "Book of Morality" by Jacques Legrand, available at Chantilly Musée Condé.
This object in the sky was nothing more than the "orb" global symbol of tripartite temporal power, which is held in so many works of art by Christian emperors or Jesus and God the Father themselves, as a symbol of power over the whole created universe.

2 / CLOSE ENCOUNTER CLASSIFICATION

• **Night lights (NL):** or the witnesses see, more than 150 meters apart, one or more lights that seem abnormal to them in the night sky.

• **Daytime Drives:** or the witnesses see the day, more than 150 meters away, an unidentified object that has not necessarily the form of a disc or saucer.

• **Radar-Visual (RV):** UFO is observed visually and appears on one or more radar screens.

• **Close encounter of the first kind** or the witnesses see a UFO, whatever it is, within 150 meters.

• **Close encounter of the second kind** UFO leaves physical evidence, such as ground traces.

• **Close encounter of the third kind** or the witnesses see a UFO and its occupants, or only pretended occupants of a UFO without it.

.Close encounter of the fourth kind

Removal aboard the UFO often including medical examinations on an operating table.

3 / CHRONOLOGY OF MAIN EVENT OF UFO SIGHTINGS

3.1 / Close encounter of the first kind

• **1492** October 11 - Christopher Columbus diary of - 5 hours before the discovery of the New World - While patrolling on the bridge of Santa Maria at about 22 hours October 11, 1492, Columbus thought he saw "a shining light a great distance.
"He quickly called Pedro Gutierrez which also saw the light. Soon after, she disappeared and reappeared several times during the night, each time dancing up and down" suddenly shines and sparkles like a flickering wick 'a candle".
Probably an atmospheric or celestial phenomenon.

• **1561** - Nuremberg, Germany: "swarm" of UFOs above the city of Nuremberg.

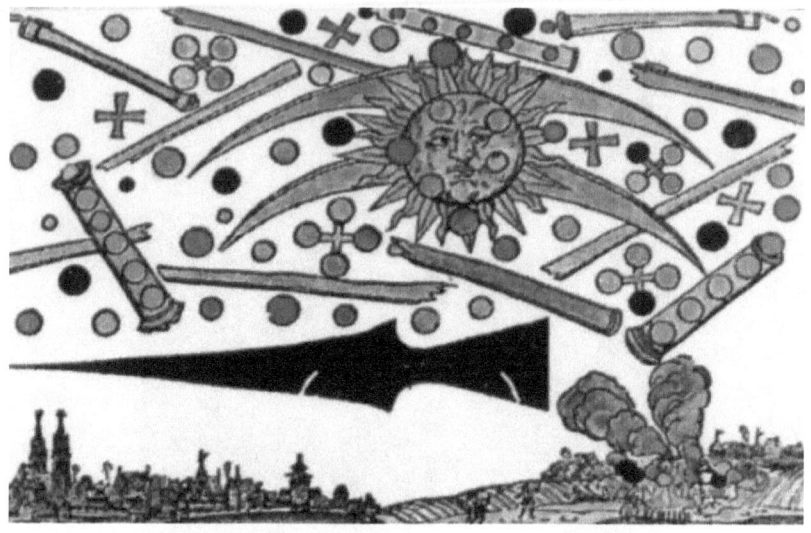

Many observations of celestial miracles are regarded as the meteorological perspective, the most common explanation being halos, mock suns, solar eclipses, lunar eclipses, auroras, ball lightning and shooting stars.

The Nuremberg document 1561 is very halo probably linked to a parhelion. According to Frank Johnson this is corroborated by the presence of two crescent moons that were placed behind the sun for lack of space in the illustration, and the fact that the event lasted an hour and did no noise. The large black iron spear may be a phenomenon due to crepuscular rays.

• **1566** August 7 - Basel in Switzerland.
Black globes invade the skies over the city of Basel. In meteorology, it is envisaged various phenomena such as a halo. The event and its description are compared to a

similar event and a description in Nuremberg in 1561.

• **1913** and after - Mountain Brow Lights; It is possible that this is a rare sighting called ball lightning.

• **1942** February 24 and 25 - Air Raid on the west coast of the United States. Unidentified aerial objects incite the military to draw thousands of anti-aircraft shells into the sky and raise the alert status of the war. That would actually come from stratospheric balloons.

• **1944** - Ghosts Hunters. colored spheres regularly spotted by military air crews around the world.

It raised the possibility of rare electrical phenomena (like the fires of St. Elmo or ball lightning).
Doctors have referred to "mirage" due to retinal

persistence after exposure to tracer shells.
The Battle of Los Angeles provides an explanation to the phenomenon where light points around the object could be reflections or splinters of the flak.

A picture of the phenomenon was published in the Los Angeles Times February 26 1942. It sees that looks like an object identified by the headlights of the DCA.
Once the Cold War ended, the archives concerning military prototypes were unveiled.

• **1946** - ghost rockets, objects with cruise missiles characteristics observed repeatedly throughout

Scandinavia.

• **1947** June 24 - Observation of Kenneth Arnold, the incident that gave the name of flying saucer.

He fidgeted in fact Ho-IX of German technology with a maximum speed of 600+ mph.

- **1948** - Green fireballs have been reported on several military bases of the United States. An official investigation followed. Can be plasma or another.

- **1952** - Case Carson Sink in western Nevada, USA. Surely secret craft.

- **1952** - Flap Washington series of observations with radar contacts in Washington, DC. This resulted in the establishment of the Robertson Panel by the CIA. Again, probably testing of military equipment top secret.

From the LOOK Flying Saucer Special Issue, 1967

The day the saucers visited Washington, D.C.

- **1953** August 5 and 6 - UFO Cases Ellsworth A UFO is seen as a glowing light and is seen by forty-five people in Bismarck, North Dakota. Possible military intelligence machinery.

- **1966** - Hunt UFOs Portage County. Several Ohio police pursuing a UFO for 30 minutes. Undoubtedly secret gear in office.

- **1979** March 5 - Observation of UFOs in the Canary Islands.

It was actually a Poseidon missile conducted by the US military.
In the picture below we can see the similarities between the observation of 79 and a rocket SpaceX for example.

• **1981 - 1986 -** UFO sightings in the Hudson Valley, where thousands of objects reports the like have been reported. The first sighting was made by a retired policeman in Kent, New York, at the end of New Year's Eve 1981. The incident at Indian Point Energy Center, a nuclear power plant in 1984.
Again it was just a test of military gear.

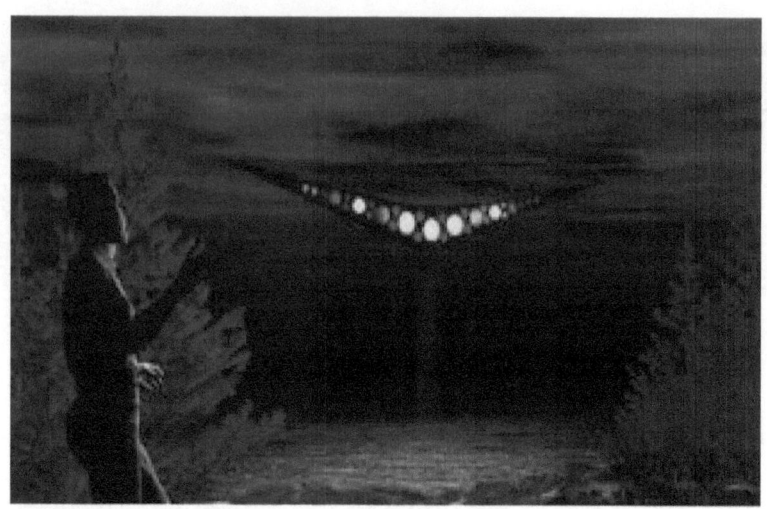

• **1990,** November 5 at 19.00 - Me and thousands of witnesses in France and other European countries have observed a huge set of lights across the sky in silence. After a few days, the event was officially "explained" by the re-entry of a rocket!

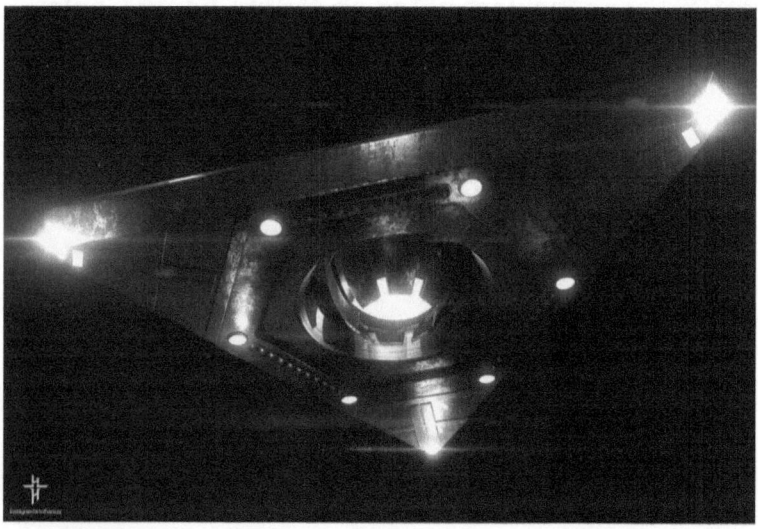

In my case, this evening of 5 November, I was in the

presence of my mother and my brother when a huge triangular object was found literally above us.
We went out every three of a taxi that took us home after our car is fell out of power an hour before (the garage will reveal the battery was completely empty!).
The ship is hovered about two minutes, during which time my mother and I were paralyzed by the UFO.
My brother had gone to hide in the courtyard of the house, he was scared.
I remember shouting in the street a 'come and see!' 'To potential people located in the perimeter, but that night no one answered in this deserted street of Lagny sur Marne.
The object then gained altitude and headed eastward and pass the sound barrier shortly after his departure.
Upon entering the house, they hesitated between the "extraterrestrial" hypothesis or a secret military aircraft project, with a preference for the first theory.
Twenty years later, I realized that it was probably the TR3B, a secret American device.

• **1997 -** The lights of Phoenix in Arizona.

It was actually the TR6 TELOS. A plane that folds!

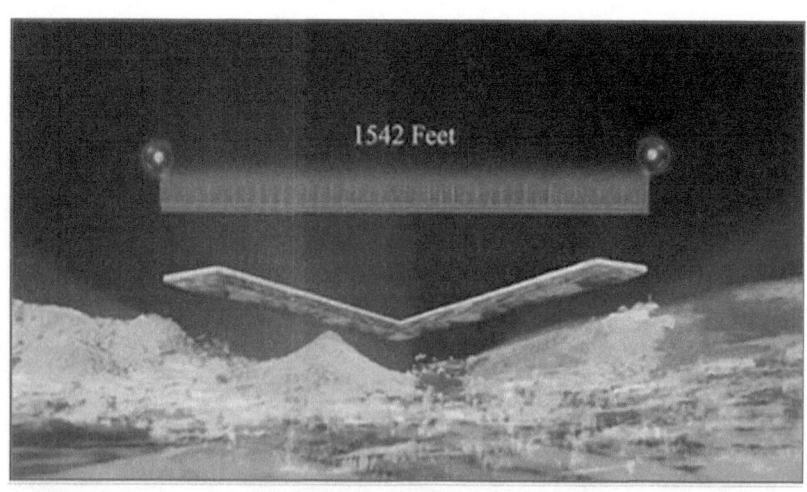

3.2 / Close encounter of the second kind

• **1947** June - Incident on the island Maury, controversial story in which a dog is killed, a boy injured by UFOs and the witness threatened by a man in black.
Six decades later, it is known that the alleged sighting was invented from scratch by Fred Crisman and his sidekick Harold Dahl, to allow Crisman write in the magazine Amazing Stories of Ray Palmer.

The son of Dahl, Charles, interviewed by Kalani Hanokano, a former director of MUFON, said that the incident had never occurred and that he himself had never been burned. The thing was confirmed to a reporter " Seattle Post Intelligencer " by his sister, Louise Bakotich, who declared never to have heard of the death of their dog on board or the injury allegedly caused to her brother.
Fred Crisman, author to his hours of fantastic stories, was inspired by the Kenneth Arnold sighting and the Roswell incident to scaffold the Maury Island and was backdated, three days before the observation Arnold.

• **1947** June 4 - Roswell Incident.
The story has become a widespread myth. Ultimately, the government has concealed something, but not an alien spacecraft.
It actually was a secret government program 40s, Project Mogul.

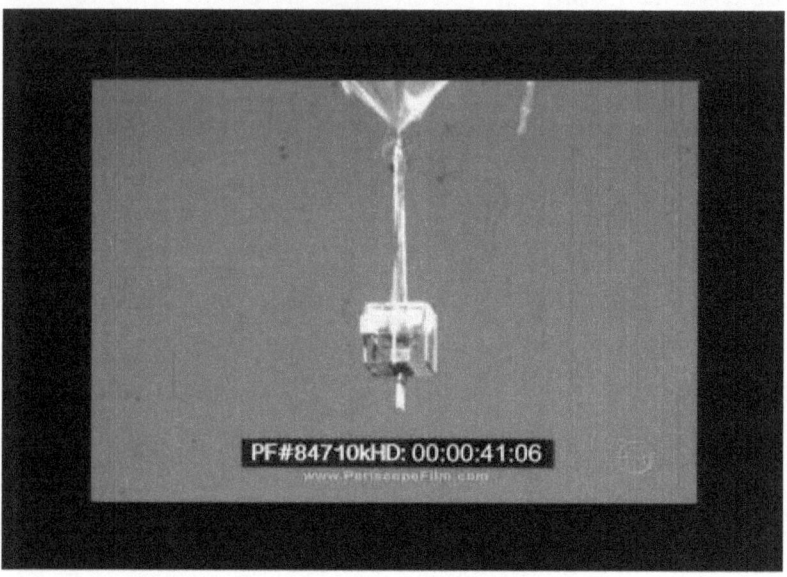

In the summer of 1947, the Russians had not yet exploded their first atomic bomb, but it was clear that this test was imminent.
He was of the utmost importance for America to know when the test took place.
Project Mogul was an attempt to listen for the explosion throwing low microphones high altitude frequency, where sound waves can propagate worldwide. Microphones, radar tracking reflectors and other devices were sent weather balloons long trains to listen to the atomic explosion.

These balloons trains were launched in New Mexico from a point approximately 160 km west of Roswell.
The Flight 4 was launched June 4, 1947 and was followed up to 17 miles from where Brazel found the wreckage, when contact was lost. The debris found in Roswell match the materials used in the balloons trains.
The Project Mogul remained secret until 1994, when Steven Schiff, Congressman from New Mexico insisted that all necessary research to be done to reassure the public and to reassure the public: the government did not conceal Roswell.
If the truth had been revealed on the Mogul project 1947, he would almost certainly have ended the speculation in Roswell debris, but the truth was revealed 50 years too late.
For many UFO enthusiasts, government secret Project Mogul has only reinforced their belief that the government also concealed the much more delicate question of contact with extraterrestrials. The Russians made their first atomic test in August 1949, which quickly became known.

• **1948** - Mantell Incident, the US Air Force sent a fighter pilot check a UFO sighting. The pilot crashed. Some claimed that he was shot.

The official investigation that followed the accident should conclude that Captain Mantell lost consciousness as a result of failure to supply oxygen at an altitude of 7500 meters and the unit then fell dive to the point impact. The US Air Force stated, first, that the pilots of the squadron and the many witnesses had mistaken the planet Venus (actually visible in broad daylight at this time) with a UFO. Then, the investigators concluded that it was a stratospheric balloon Skyhook Navy released from Clinton, Ohio.

• **1948 -** The incident Kapustin Yar, where a shaped UFO "cigar" was shot down by a Russian MiG.

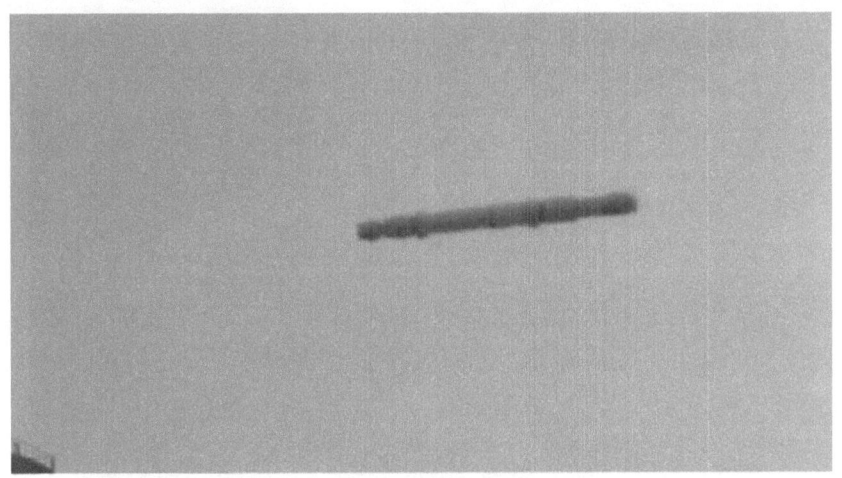

Kapustin Yar was the most sensitive air base in the former Soviet Union, even exceeding the area 51 of America because of the level of secrecy that covered. Kapustin Yar was created to accommodate the development of the space program of the Soviet Union after the end of World War II. It is located more than 500 km south of Moscow and about 60 km east of Volgograd. It was here that the V2 rockets were captured and that German scientists who created them were ready to work with the sole task of entering the space before the Americans, but also to design and test new aircraft.

• **1950** , August 15 - Mariana UFO Incident in Great Falls, Montana.
Nick Mariana, general manager of the baseball team minor league Great Falls Electrics, and his secretary, Virginia Raunig, inspected the empty baseball field from Legion Stadium before the game, when a bright flash caught the eye of Mariana. They would have seen two shining silver objects, spinning flying over Great Falls at a speed he considered to be 200 to 400 km per hour.

He thought they were about fifty feet wide and one hundred and fifty feet away. Mariana ran to his car to get his 16 mm camera and filmed the UFOs for sixteen seconds.
It was probably a military test (device or plasma).

• **1953** - Felix Moncla, army pilot, US Air, disappears chasing a UFO.

Moncla has either hit a secret device, or the victim of a lack of oxygen leading his plane down in a lake in the flight area.

• **1976** September 19 - The Tehran incident. A UFO over Tehran, Iran has neutralized two electronic components of the F-4 interceptor and ground controller.

Probably a secret anti gravity or a camouflaged tank unit.

• **1978** - Kaikoura Lights, a series of observations by a cargo plane "Safe Air " flying off the coast of Kaikoura, New Zealand, was escorted by strange lights of colors and varying sizes.

Maybe due to a plasma test or holographic projection.

• **1980** December 29 - Cash-Landrum Incident. A diamond-shaped UFO irradiates three witnesses near Dayton, Texas. They had to be treated for poisoning by radiation and have sued the US government to court, military helicopters have been seen with the UFO.
The victims demanded 20 million dollars in US government hired a trial. The object being escorted by US helicopters, Betty Cash (50) and Vickie Landrum (60) declared to be victims with a child of seven years of military experience.

29 Décembre 1980
Engin lumineux en forme de diamant aperçue
par Betty Cash et Vickie Landrum près de Huffman
(Texas), entouré d'hélicoptères.

If the reality of their radiation was scientifically proven, no evidence for the presence of the helicopters could not be provided, and the US military denied obviously have gear like the one that had irradiated. The trial was therefore lost by victims in 1986.
Probably therefore a project / secret military test.

• **1986 -** Brazil, more than 20 UFOs are seen and picked up by radar in various parts of the country. Colonel Ozires Silva, former Embraer and then president of Petrobras, flying an airplane Xingu PT-MBZ near the square, is informed and tries to continue UFOs.
Two Northrop F-5 and Mirage III three are eventually sent to intercept them. After watching radar and visual objects have disappeared.

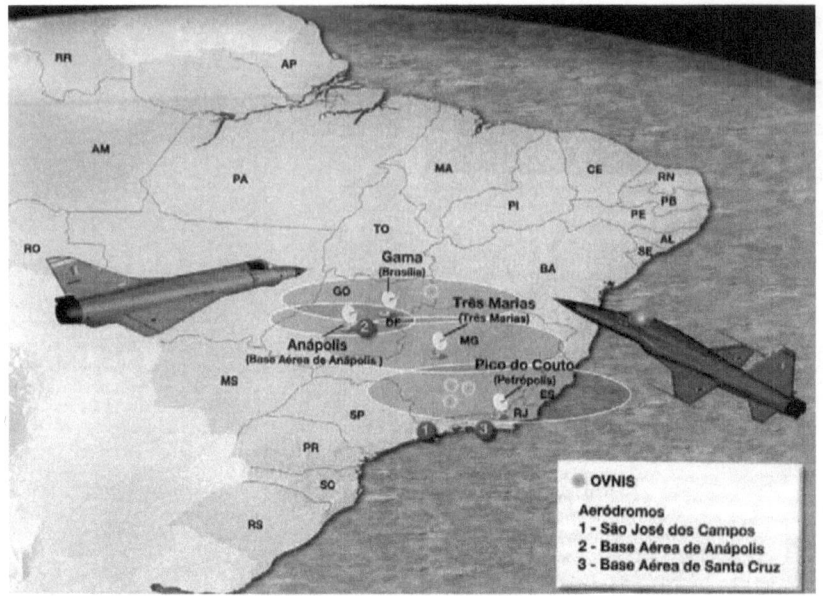

For this case: Possible plasma shots.

• **2003** - UFOs have been sighted before the 2003 blackout in North America. Certainly anti gravity military devices emitting electromagnetic waves resulting in the failure.

3.3 / Close encounter of the third kind

• **1946** May 18 - Gösta Karlsson claimed to have seen a UFO landing near Angelholm, Sweden. On the site, a model of flying saucer was erected.

Clas Svahn, chairman of UFO-Sweden, investigated the case and wrote a book with Gösta Carlsson on the incident. According to him, there was no convincing evidence that the event took place as described by Gösta Carlsson.

• **1955 -** Kelly-Hopkinsville encounter
A group of strange and playful creatures who intimidate a family while she shoots them.
It was envisaged that the family was able to see the great owls Dukes.

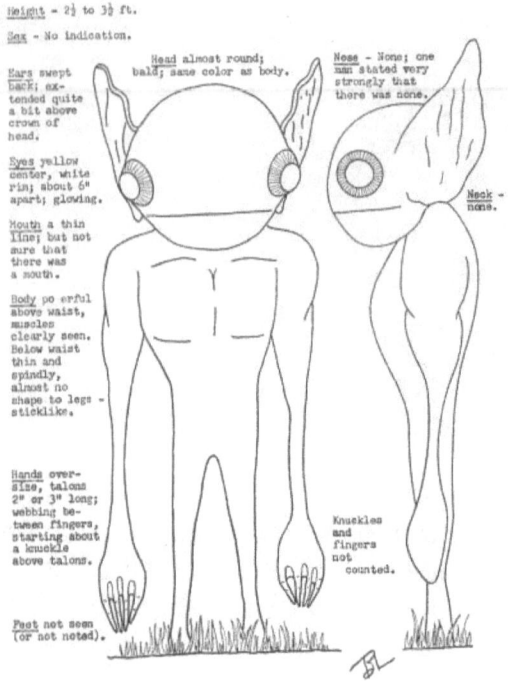

I personally advance the thesis of the bat chasing around the house. But many other things are possible.

• **1961** - Hill Abduction, the first abduction experience by widely aliens.
The Hill may have been the first victims of MKULTRA and MILABS program namely beings put under hypnosis and

drugs.

• **1964** April 24th - Lonnie Zamora, a Socorro police officer in New Mexico (United States), tells of a close encounter. Maybe a secret military balloon.

- **1973** - Pascagoula, an experience of " removal " by and well known.
Can be an operation of MLABS.

- **1975** - Abduction Travis Walton, a logger from Arizona who was allegedly abducted by a UFO on the side of a road. At the scene of the incident, no trace has been found.
Philip Klass think removing Travis Walton was a clever staging designed by Michael Rogers, employer Travis Walton.

He was unable to honor the contract for which it was committed. The deadline was approaching and he still had several hectares unclear.
You should know that in the US contract may be terminated if the work is carried out in unacceptable security conditions.
The alleged abduction therefore provided to Michael Rogers legal grounds for breaking the contract. According to one of the psychiatrists who examined Travis Walton,

Dr. Rosenbaum, Travis Walton was not removed but have had an experience related to a particular psychosis. (Hypnosis / MILABS?) Also, The Walton sold their story to the National Enquirer.

• **1978** May 10 - Removal of Emilcin - a man Emilcin in Poland would be abducted by gray. On the site, where it should have taken place, there is now a memorial. Wolski's story was made public by a Wawrzonek Witold, a character involved in false stories of UFOs.

• **1978** June 19 - a family driving through Oxfordshire, United Kingdom, is intercepted by a UFO. Can be a PSYOP (helicopter/ drug / hypnosis).

• **1981** January 8 - Case Trans-en-Provence, mutual UFO landing.

The skeptic Michel Figuet assumed that the witness had seen a helicopter and had not recognized due to the reduced visibility in the twilight of winter; the and traces the effects observed on the ground may have been caused by the passage of a cement mixer.

• **1989** November 30 - Linda Napolitano was abducted by a UFO. Witnesses would include the UN Secretary General Javier Perez de Cuellar and Guerra, as well as CIA agents. Budd Hopkins investigated the incident report. It was probably a MILABS operation (hypnosis / Drug).

• **1994** September 14, 1994 - A UFO file through the skies of southern Africa. Two days later, " something " landed in a school yard in Ruwa, Zimbabwe.

It would have been viewed by 62 students.
Cynthia Hind, a pro-UFO writer, interviewed them after the meeting and made them draw pictures of what they had seen. This case is considered one of the best proofs of ufology, is often mentioned in the "blogosphere" UFO.

It is important to cut this case in two stages that are important:

The coming and Cynthia Hind interview sessions and those of John Mack, two months later.
Mack then close interest in the abductions, but also ecology and anti-nuclear activism.
Also, one may wonder if, because of this strong cultural background investigators, it would not have contaminated the evidence.
It is revealing to note how the protection and ecology of the planet appear in children's stories when Mack questioned, while this theme was not present in the early narratives collected by Cynthia Hind.

It is important to know that if Cynthia Hind moved to school is because there was a favorable and exceptional context: the country was under a "UFO flap" September 14 1994. But what then was this object that flew over the sky and resulted in reports of UFOs: a vehicle came from another world and who would later landed near the school or something much more conventional, though surprising to observers?

Cynthia Hind et John Mack

These UFO reports of September 14, 1994, have just been harvested and were the subject of a Cynthia Hind's article in the journal to which it has contributed a lot, " UFO Afrinews ". This article indicates that the observations of this object spinning in the sky have occurred between 9:05 p.m. 8:50 p.m. ET. Some of the witnesses produced drawings of the UFO.

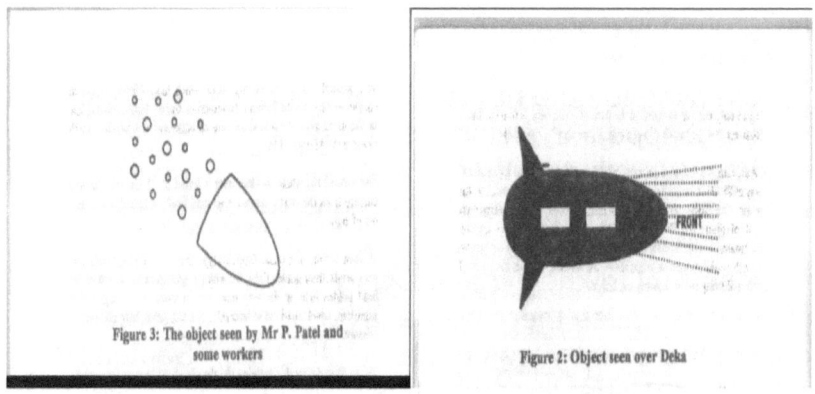

Figure 3: The object seen by Mr P. Patel and some workers

Figure 2: Object seen over Deka

Here are a few : Southern Africa, especially around Harare, she experienced the passage of a visible reentry from its soil? The answer is yes and is in the catalog Cash Atmospheric Ted Molczan.
The time also corresponds (6:51 p.m. 8:51 p.m. UTC = local).
From software, it is even possible to know whether the back was visible example to Harare.

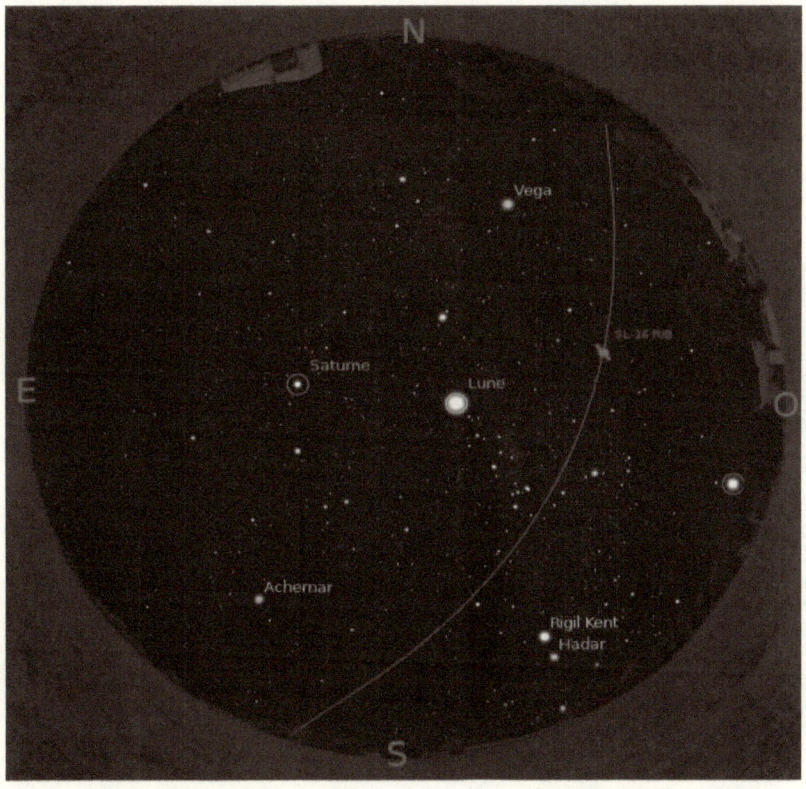

Also, the object that flew over the skies of Southern Africa and gave legitimately (and classically) result in UFO reports was a re-entry.
It is during this context that Cynthia Hind went to Ariel School to talk with students. The interview methodology

with children Cynthia Hind is very far from these standards so at least control these biases and confounders (especially in young children), but it maximizes the worst.

Criminologists and psychologists already recommended two main methodological criteria regarding the interview with the child Cynthia Hind fails to comply with the individual interview and the fact of promoting the free narrative. In this case, the free narrative is not respected: The interviewer, the investigating officer (or not) or practitioner should always interview the child on his memory and his recall of the event, leaving the child to deliver his testimony without being interrupted nor questioned.

It is in the second time that we can initiate a second phase to maintenance, the questions, but again with a number of standard precautions and procedures required. In the case of the school of Ariel, maintenance sessions are collective.

Children are interviewed in "rank" of four to six, sometimes other children are present and listen to another questioned child. As the children hear what others say, at risk of becoming affected.

One can also ask whether the fact that the sessions of drawings that took place at the school, were not conducted and give the same kind of performances every time.
A student (S. Charity) Tim Leach interviewed by the BBC said:
" I have not heard of UFOs before we talk about with adults, journalist and parents. " Another child (Lisa P.) says he first thought of an "alien" before she realized that it could not be that of the gardener.
When were they made? It's really not clear and this time is a very important variable in view of all the psycho-socio-cultural factors mentioned throughout this article and having potentially impacted these drawings.

According to Cynthia Hind in UFO AFRINEWS, drawings were made at the request of HeadMaster when children are back in school.
But another source written by itself, it would be she who asked that children draw before his arrival, two days after the "incident."
Finally, at this conference the director of the future documentary speaks of drawing made one or two days later. In addition, there are two drawings sessions, the demand Cynthia Hind (one or two days after the event) and that of John Mack (two months later).
Often, designs of one and the other of the two sessions are mixed in sensationalist presentations (as in the conference link above).
How many drawings? The UFO media suggest that was 62 drawings describing the same thing.

But Cynthia Hind itself speaks from 30 to 40 drawings, and says she has 22 of the most explicit and clear. They are almost always the same as presented in the UFO issues or items.

In this regard, it is important to note that the drawings are mixed performed at the request of Cynthia Hind and those made during the visit of John Mack, two months! Had do we selected them because they were more sensational "alien" UFO?

Note that some of the witnesses doubt the allegations of others or even confessed having "lied". Shelley S. For example:
" I heard later that Lizel and another guy whose name I have forgotten admitted they lied to get to be on TV and my sister and I were there and we looked and could not see anything, apart from the fact that where everybody was pointing his finger on something that was of kilometers from us, how someone would see such details, even if something had landed there? Note that no adult is involved in observing or has seen anything.

Moreover, Mack himself wrote that the children ran excitedly returning to school to tell teachers, who were in a meeting, what had happened, but teachers were initially rejected this as being the imagination of children, and when they finally went to, there was
nothing.

The Coming of Dr. John Mack

Mack did not go to school on the spot as Cynthia Hind, but two months later.
Before his talks with the children, he will intervene and respond to at least interviewed on Radio702, during a TV

program that followed national news.

On 30 November 1994, it will make a speech to local Sport Club with 300 guests and present its thematic and it will be many interactions with the school principal, teachers and parents.
Then he will ask the children to school. It is found in the method of J.Mack, the trend acquiescence namely an attraction of the subject to positive responses and more readily respond yes than no.
This is why psychologists and criminologists have established strict protocols and standards of maintenance, especially with children. It was shown on issues the following:

The child is never forced to imagine in the question.
For example Mack often asks the children: " Tell me what you think is the reason for the alien visiting Earth? "
This kind of question requires the child to imagine and find an answer quickly because he knows that the adult waits.

John Mack:
" Why do you think they want us to be afraid? "
The Child: "Maybe because we do not take care of the planet and the air properly."

The children here have probably guessed the desired response, one that would have or want to hear the Dr. following themes dear to him. Let us quote this example from John Mack and a child. Two multiple-choice questions, the child chooses twice the second alternative, even contradict themselves.
John Mack: (first alternative) " Is it an idea you had before, that we do not properly care for the planet and take air or (2nd alternative), this idea has come to you when you you had this experience? "

The child: ". When I had this experience " (She chose 2)
John Mack: " At what point did you feel that? (1) When you saw the craft, or (2) when you're back home in the evening? "
Child: When I came back.
(She chose the second alternative everytime, even contradicting his previous answer ...)

So this case is for me a perfect example of a long process of interaction and psychosocial variables on real background of luminous phenomena observations added to other terrestrial elements as a possible observation of military gear or not.

• **1996** - Varginha incident; a strange creature is seen in Minas Gerais, Brazil.
In 2010, extracts from official military investigations conclusions on the matter were published by the magazine ISTOÉ.

They have caused great anger among the enthusiastic because Lieutenant Colonel Lucio Pereira Finholdt raised that the most likely hypothesis is that a disabled resident of the neighborhood called Little Luizinho or Luis was taken to an alien on the evening of observation.

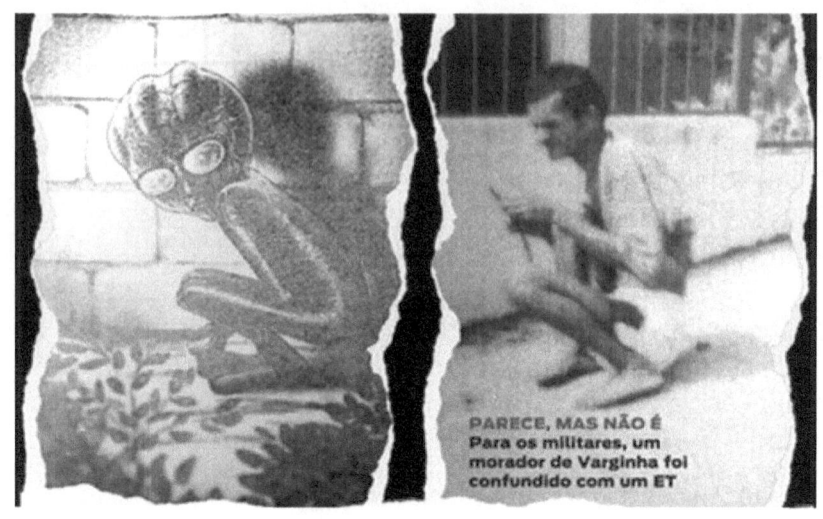

PARECE, MAS NÃO É
Para os militares, um morador de Varginha foi confundido com um ET

4 / "ALIEN" ABDUCTION & MILABS

It seems that the phenomenon of alien abduction said often serve to cover the reality of operations carried out by many human beings; The government. It there's several projects such as MILAB as MKULTRA and Monarch. They are called MILABS, abductees,
who are kidnapped, trained and used as deep blacks agents by elements of military intelligence for various purposes. These deceptive operations are carried out in order to convince the target that they encountered aliens, which is actually a ploy staged.
DARPA is the organization that maintains the MILAB projects.
Most kidnappings and injections are administered by external physicians, away from main operations used by many organizations.

The mind control, intimidation and torture are used as a way to control the person. Over time, it breaks people's memory MILABS.
The projects are operated in military bases, one of these bases is the Wright-Patterson Air Force Base in Dayton, Ohio.
Some MILABS / MKULTRA infiltrated and used throughout the Western world to make dark operations.
I think as well as many others, these projects also hide child pornography and other abuses.
Here are some quotes and testimonials on this subject:

The journalist Alain Gossens died in 2010 wrote in 2007:

"In recent years, a growing number of authors and American researchers like Helmut Lammer, Karla Turner and Walter Bowart have drawn parallels between the victims of the three seemingly distinct phenomena:
The ritual abuse victims, those who report being abducted

by alien entities and guinea pigs for behavior control experiments perpetrated by psychiatrists in a militarized context.
Some victims even belong to three categories the point that it is difficult for them to understand the content of the experience extraordinary and traumatic which they were subject. The stories collected eg by specialist about alien abduction that was Karla Turner, now deceased, show how it is sometimes difficult to make a distinction.
In all three cases (ritual abuse, abductions and Mind-Control), you must understand that consciousness, memory and therefore the perception of the world are deeply affected by the violence of the experience and its strangeness.

Another common point between these three types of events:

The victims are almost always affected in their sexuality. This is obvious for ritual abuse because it is often the same purpose practices: submission obtained by ritualized rape. In the case of alien abduction, the victims almost always suggest that the experiences which the alien entities submitted them had their genitals to target: egg or sperm samples, penetration of private parts by sensors, artificial stimulation of sexual activity through an unknown technology.
When the Mind-Control experiments also speak some victims of sexual abuse, rape, etc.

There is still something in common between these three phenomena:

How memories come to consciousness.
In most cases, the ritual abuse victims of alien abduction and Mind-Control will remember their traumatic

experiences as part of a therapy.
Memories, over hypnotic regressions, then arise unexpectedly, totally disjointed and it is with the time and effort that these memories are then ordered to form a coherent whole with a semblance of chronology. It is during this stage that some therapists are trying to validate the testimonies trying to cross them with hardware, the other evidence victims, etc.
The memory that is at the center of these manipulations is deeply affected, but this is not to say that all the memories are not reliable.
The whole problem is to know what is behind these memories. And above all, why human beings are subjected to such treatment. The phenomenon of alien abduction maintains much deeper links you
might think with the Mind-Control and abuse and satanic rituals.
All these phenomena are "working" on consciousness, memory and perception of reality.

Dr. Richard Boylan who has written extensively on the subject of aliens and UFOs, met and interviewed many witnesses.
He found five common among people who were abducted or have seen aliens or UFOs:
- These individuals possess a high level of psychic abilities.
- Similar phenomena are observed with other family members.
- Children who have suffered severe abuse or trauma.
- Individuals or entire families linked to the government and /or intelligence agencies.
- They were often American Indians, native. There is also a strong correlation between the sites are practiced occult activities such as ritual abuse, secret military facilities and UFO events and testimonies of abductions (abductions)

by ET In his book "Satanic Ritual Abuse , Principle of Treatment ", Dr. Colin Ross explains the
strong similarities between satanic ritual abuse and abductions:
"There are thousands of people in North America today who have memories that go back about an alien abduction in spaceship, with experiments performed on them (...).
These "abductees" arrive in therapy with periods of missing timeas well as post-traumatic symptoms inexplicable, like satanic ritual abuse survivors.
Abductees report that they had amnesia barriers hypnotic deliberately implanted by AND, the survivors of satanic cults describe exactly the same programming made by their tormentors.
The survivors of abuse satanic rituals also describe forced pregnancy, medical experiments in laboratories and abortions before the end of pregnancy. "
In his book "Mind-Control World Control" Jim Keith writes that the extraterrestrial abductions business would cover experiments on mind control actually performed by humans in the flesh.

Excerpt from an interview with Kathleen Sullivan

Kathleen Sullivan at 16, and today:

"Can you talk a little bit of memories related to the phenomenon "UFO "?
I can only share with you what I have personally experienced.
What I will say may match or not with what other survivors can remember.
One method that has been widely used to trick my mind, even as an adult, was Ericksonian hypnosis.
This worked well because I almost always was in a trance, anyway. It was very easy for a gifted person in hypnosis, say certain words or phrases to influence me and make me believe such a hovering helicopter was a UFO and its occupants were aliens from Uranus.

However, I remembered these experiences, although I thought the helicopter was a UFO, I actually saw a military co-pilot. I had a strange experience in which I was taken to an underground facility with large tunnels. I was presented by adults dressed in white suits,
to what I was told to children "aliens". Although at that time I thought they were aliens, later a handler told me they were the result of decades of genetic experiments. He also explained that these children are raised to

believe that they are of extraterrestrial origin and even receive education in a very different language. The
On another occasion, I was escorted by the individual I call "Lucian" in a large warehouse room that had been configured to look like the inside of a UFO.

We entered through an airlock of the size of a shower consisting of a bright and soft silvery metallic substance. In this UFO "cardboard", a number of adults wore costumes clearly "aliens". Lucian boasted, as usual, that other people like me had been taken in this particular installation is to receive a type of programming abduction UFO / alien. From that I had not been hypnotized to believe that this place was inside a UFO and Lucian had needed to boast of his intelligence, I never forgot that it was there a setting in place.
I believe today that these types of false memories are created in the minds of slaves to discredit them if they ever remember the "aliens" and they start talking. I also think that these false memories are installed to block the memories of the real perpetrators and the actual traumatic experiences. "

Testimony of Cathy O'Brien

"Bennett Johnston submitted myself to other mental manipulation that involved not occultism but the theme of aliens. These guys who handled my mind and me were programming for MK, real criminals in control of our country, claimed to be gods, demons, aliens ... This in order that I feel totally helpless, I integrates the fact that they were still there behind me to hurt me. And it worked very well at that time ...
Bennett Johnston stated that he was an alien. He told me that he participated in the "Philadelphia Experiment" and when the ship was gone, he returned to spacecraft ... It joins the theme of "mirror air / water" frequently used by NASA is an inversion / inversion.
Because, again, the subconscious mind has no reasoning ability.

Bennett Johnston then showed me on the General Dynamics website, a stealth craft "top secret".
It was a triangular thing that was not in any textbook, which no one spoke, which was not visible in the newspapers, but who was nevertheless there, suspended

in the air before my eyes ...
It was another one of these systems of the army top secret. For me at that time, it had me look like a spaceship! I had never seen anything like it. Thus, everything was Bennett Johnston was then
for me in connection with extra-terrestrial. It was so easy to make me accept that everything that was happening was actually perpetrated by aliens. I do not say that aliens do not exist, it would be stupid of me, but
what I mean is that they are people who actually claim to be ET.
If there is a reality for an extra-planetary influence, we need to clear up misinformation and manipulation of minds that our government practice.
I know for a fact that their plan is that we all felt helpless ...because under a so-called alien domination and that our "Independence Day" is preparing ... So beware of that! Understand that these criminals confiscate us the information and technology under cover of "National Security".

They have at least 25 years of technological advance on us all! Can you imagine what they have today?

What happened in the last 25 years?

The microwave ovens, computers, but they continue on their side progress and they have a lot in advance. So when they say "It's from aliens!", Showing us an incredible technology, do not fall into this trap you feel totally helpless. Superstition begins where knowledge stops, and we were isolated from this knowledge for a long time. People have always had different beliefs and I'm sure all of you have different belief systems. Whatever your belief system, it is imperative that you know that these criminals are human, they among us to harm us.

They must be held accountable for their actions and their crimes against humanity. "
Thus these various testimonies confirm the idea that indeed all the victims of so-called alien abductions are actually victims of governments and / or secret organizations. Ultra power drugs have been and are still used as ketamine, LSD or psilocybin.

Most of the cases discussed in this book comes from the USA Would there but a lot to say about pedophilia and kidnapping in France. I think for example in the case of small Amoris, son of Severine Ravat which tells in a video he undergoes tests in the presence of an alien in a military base in Lyon. It has actually been drugged and sexually abused at a military base in Lyon by his father and other well-placed characters and often Freemasons. By putting myself in the place of those people who have suffered so much by the abuse, I can understand that all do not want to do a reverse hypnosis work to understand what really happened to them. Maybe they believe had surgery, inspect, manipulate see raped by an alien may be more reassuring, exotic and / or fascinating than to know that was actually done by a sexual abuse sick or perverse. All in the name of power and money.

5 / "ALIENS" IMPLANTS

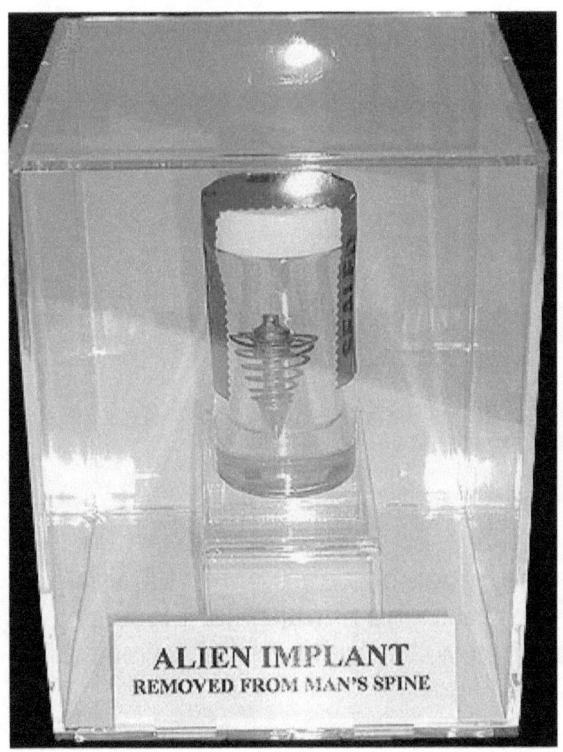

An argument that is often among people who believe in aliens is that implants that find themselves in some contacted or abducted.
Yet again a simple explanation is necessary;
- Either the person was a victim of the shadow government (CIA ...) and it was inserted an RFID chip (invented in 1973).

- Either the person is lying and it is a hoax. Three cases of implants:

1 / Case February 2005

A man had a sore nose, when he came out of the living room the next morning, he found a triangular piece of cloth on the floor, just below the spot where he was sitting yesterday.

Analysis Conclusion

The object found on the ground is identified as a synthetic polymer. Specifically, the polymer composition is identified as polycarbonate + poly (styrene: acrylonitrile: butadiene).

This shows that the object is not a foreign implant.

2 / Case of May 21, 2005

A lady, who claims to have been abducted by an alien found a small chip in his mouth.
There is speculation that it could be an implant.

Analysis Conclusion

The object is certainly not a foreign implant. It is clearly identified as 7 ½ lead fired from a shotgun shell. Properties such as diameter, weight and softness correspond to those of the reference lead shot.
In addition, the polydimethylsiloxane is detected on both surfaces. This material is common and used in the manufacture of the blow.

In addition, lead carbonate is indicated for both sample Without more background information, it is not possible to determine how the shot shotgun resulted in the mouth of the witness. She may have eaten an animal killed by a shotgun.
The 7½ size is generally used for smaller game such as duck, grouse, dove and quail.

3 / Case of March 11, 2006

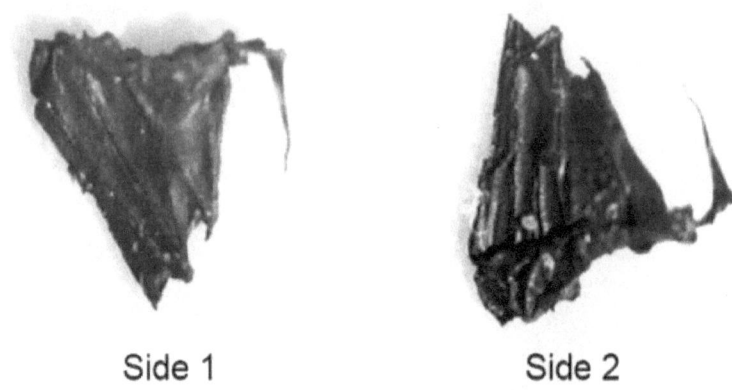

Side 1   Side 2

On 11 March 2006, the witness was in the shower around 11 am
45, when a substance he fell nose.

Analysis Conclusion

The substance is identified as a synthetic polymer. Specifically, the composition is a poly (styrene: acrylate ester).
This is probably the burning plastic, it is not a foreign implant.

6 / ANIMAL MUTILATION

The first known case of a strange cattle mutilations occurred in Alamosa, Colorado, in 1967, a hundred kilometers north-east of Dulce. An appaloosa horse was found with head and skinned and scrawny neck.
The bones were white and clean and there was a lack of blood in the region. Lacerations were cauterized like a laser scalpel was used, according to a pathologist Denver. No satisfactory explanation has ever been found on how or why this animal was killed.
Since then, hundreds of cattle mutilations were discovered. The ranch with the largest number of mysterious animal deaths belongs to the Gomez family near Dulce, in the Valdez area.

Unlike the breeder whom I spoke, Valdez said he had never seen anything that took him to believe that there were aliens in New Mexico, even claiming to have found equipment military such as gas masks and glow sticks

around the carcasses of mutilated cattle!
Valdez also said that UFOs seen in the area were actually military vehicles advanced, some of whom were involved in the capture of these animals in order to conduct experiments and then return them without being seen. He said the animals were mutilated in order to convince the people that it was done by aliens. Regarding the identity and motivations of this secret military group, Valdez said these issues were too sensitive to be discussed.
Valdez is not the first to suggest the existence of a secret underground base near Dulce. In fact, the legend in UFO circles would be that there would not a secret underground base, it would also be a joint American and foreign facility.
These rumors started with two gentlemen with whom Valdez had befriended Paul Bennewitz and Richard Doty. Paul Bennewitz had a technology company located at the gates of the Kirkland Air Base.

Albuquerque, New Mexico. In 1980, he reported to the US Air Force that strange lights filmed above the base and gave signals that he thought to be extraterrestrial in nature. Instead of informing Bennewitz he watched the secret military projects, Richard Doty, officer of the air force intelligence service, was responsible for encouraging the convictions of Bennewitz on terrorism and perpetuate with fabricated evidence from scratch.
By going to convince Bennewitz there was a military secret base beneath Archuleta Mesa, which dominates the town of Dulce, New Mexico. Greg Bishop, researcher and author, wrote about Doty, and Valdez Bennewitz in his book Project Beta. He explains that the purpose of the fraud was to fight against the counter-espionage of the Cold War.
Richard Doty was ordered to lie to Paul Bennewitz to conceal the US military secrets he thought too sensitive

to talk about. All the while, Valdez remained friends with both men, ultimately helping Doty to get a police officer in New Mexico.

In his 1997 article entitled "Dead cows that I knew" the researcher Charles T. Oliphant speculates that the cattle mutilation is the result of secret research on emerging diseases of livestock and the possibility that they can be transmitted to man.

Oliphant says the NIH, CDC and other organizations funded by the federal government could be involved and that they are backed by the US military.

Part of his assumption is based on allegations that pharmaceutical products for human use have been found in cattle mutilated, as well as autopsies showing that cattle mutilations usually involve areas of the animal that are related to the "inputs , outputs and breeding. " To support his hypothesis, Oliphant cites the case of Reston ebolavirus in which military officers in civilian clothes traveling in unmarked vehicles, entered a research facility in Reston, Virginia, to recover and destroy secretly infected animals by a highly infectious disease.

In addition, a 2002 report NIDS reports the testimony of two police officers from the Cache County, Utah.

The region has witnessed many unusual animal deaths and ranchers organized armed patrols to monitor the unidentified aircraft which, according to them, was associated with the death of livestock.

The police witnesses claim to have met several men in an unidentified helicopter of the US Army in 1976 in a small community airport Cache County. The witnesses said that after the stormy meeting, cattle mutilations in the area stopped for about five years.

The biochemist Colm Kelleher, who investigated several alleged firsthand mutilation, argued that the mutilation is most likely a clandestine effort by the US government to track the spread of bovine spongiform encephalopathy (BSE) and related diseases such as scrapie.

The theories on government involvement in cattle mutilations were also fueled by the observations of a black helicopter near mutilation sites.

On April 8, 1979, three policemen in Dulce, New Mexico, reported a mysterious plane that resembled a helicopter of the US Army hovering around a site after a mutilation that has killed 16 cows. July 15, 1974, two unlicensed helicopters, a white helicopter and an airplane Twin black, opened fire on Robert Smith Jr. while driving his tractor on his farm in Honey Creek, Iowa.

This attack follows a series of mutilation in the region and the neighboring border of Nebraska. The reports of a helicopter's involvement have been used to explain why some cattle appear to have been "released" from a considerable height.

7 / THE CROP CIRCLES

These were observed mainly in the South of England (Hampshire, Wiltshire ...) since the 60s, and more recently in the world.
To date, their number was estimated at at least 5000.
At first, it was found flattened ears of wheat in the form of a circle.
Since then, the phenomenon has become greatly complex and diverse. Today, the motifs found in the fields reach hundreds of meters and forms complex geometric are completed in a very short time, e.g., less than half a minute. Many hypotheses have been examined, including aliens, because UFOs were often seen nearby, sometimes projecting a beam of light toward the ground.
It is likely that these geometric designs are due to the firing of a cannon military military microwave computer

controlled.
The arguments in support of this view are:

7.1 / UHF MICROWAVE

The research of Dr. Levengood (of BLT Research Team), American biophysicist, corroborated by the analysis of Ken Larsen, British biologist, showed how the stems (wheat, rapeseed ...) are flattened without being broken or damaged is typical of a microwave effect UHF.
Thus, one can see canola stalks bend to 90 degrees, whose flowers are still intact, although these rape stems break easily when attempting to bend them by hand. The new position taken by the plant becomes fixed. It continues to grow horizontally and breaks if one tries to restore the upright. Of electromagnetic origin phenomena were observed in the locations: irregular behavior compass disturbance devices electrical, radio frequency interference, flashes of light, crunchy sounds, animals obviously indisposed, dowsing effects, etc.
Many positive effects (spontaneous healings, feelings of peace...) or negative (temporary paralysis, mental confusion, memory loss, terror ...) have also been observed in humans.

Remember that some effects could also be explained by a fertilizer or pesticides reaction subjected to microwave radiation, reaction likely to release toxic gases.
The appearances of light flashes and crackles are not objective phenomena and can only be induced sensations in the brain of a witness by an electromagnetic field. Albert Budden provides an e
xample of such magnetophosphenes "If the subject's brain is exposed to a magnetic field [AC] whose frequency varies from 10 to 100 Hz and whose power varies from 200 to 1000G, the subject see flashes of light

in the upper left corner of his visual field. The investigator Busty Taylor showed that samples of plants or soil taken from a crop circle could be attracted by a simple magnet.

This could be explained by the fact that the ferromagnetic particles in the dust of the atmosphere were linked to the circle of the site, after or during its creation.
Some of these particles were examined under a microscope.

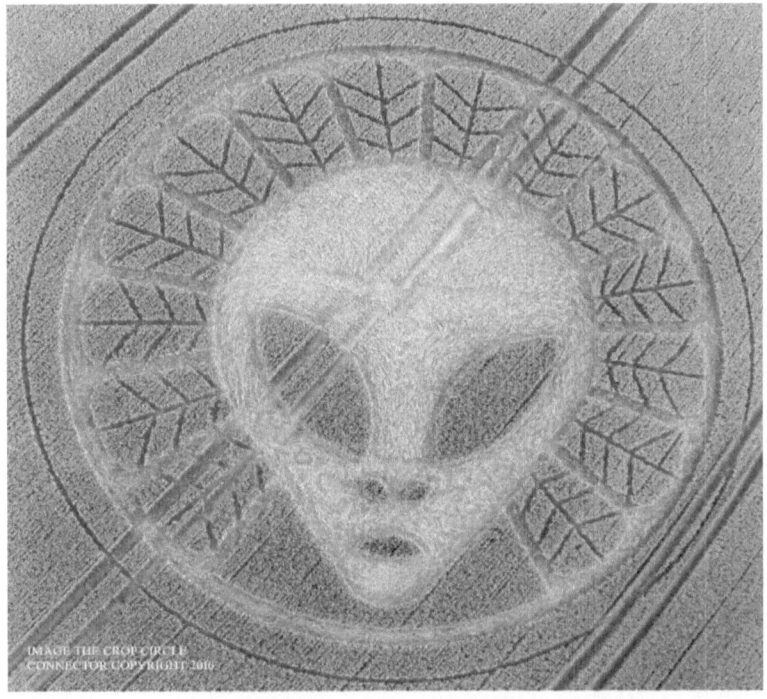

They seemed to have melted in contact with soil or plants to create a fine crackle glaze. A bird was found in special training and his body seems to have exploded, as if it had been cooked alive in a microwave oven.
In other formations, dried hedgehogs were discovered.
In addition, the wheat grains are dehydrated and crisp.

They are less conductive as the work of Dr. Levengood. Laser technology microwave appeared in the 50s and improves in parallel with the increasing complexity of crop circles. Over 50% of observed circles in England appeared cloudy or rainy: the cloud cover would hide behind the shooting microwave.
Microwaves can penetrate clouds and act through the rain, and they may be less damaging to plants when it rains.

Some patterns suggest the use of a variable diameter rotatable beam according circles considered, which could match the natural or deliberate dispersion of a laser beam shot from a high altitude.
The diameter is estimated at less than 30 centimeters at a distance of 20 km.

7.2 / 3D DESIGN

The geometric designs observed today are typical of those that can be seen on computers: 3D designs, fractal designs ... Some models are mathematically quite complex.
The designs are drawn very quickly, day and night, sometimes in front of witnesses who claim to have seen the wheat flatten before them in a few tens of seconds. Three stories of this type have been registered to date, from
controls known and trusted.

7.3 / HPM TECHNOLOGY

The HPM technology (High Power Microwave) is now used by the army to destroy the enemy electronics. "Star Wars" (SDI) President Reagan proposed the establishment of various missile laser devices.

Include only GBL (Ground-Based Laser), a gun on the ground for a satellite that sends reflecting the light beam to a combat satellite mirror, and SBL (Space-Based Laser) that directly targets the target.
While the overall project was abandoned some of its devices
may have been completed small scale.
The firing can also be performed from an aircraft (airborne laser) or a balloon positioned at a height of 20 km and stabilized for example by ion propulsion engines.
In 1991, a drawing appears outside the residence of British Prime Minister John Major, pointing to the house.

Obviously, this accommodation was under strict surveillance by fear of terrorism of the IRA. What other organization outside the secret service was behind such a design? Many crop circles have also been observed in closed military zones under monitoring.
The army wants to maintain the belief in aliens and conducts psychological warfare tests.
They have the means to attract observers with false saucers so that the circles are attributed to aliens.
The military secret services are not hampered by the risk of killing.

Unfortunately, the "circles" have already produced a victim: October 22, 1987, as the jet passed over a crop circle, the pilot ejected and then detached from his parachute before hitting the ground. Some aerial photographers reported that the drawings have a significant influence on them and on the control of the aircraft when flying over them.

The British Army is cooperating closely with the US military and may have accepted the use of "his" land. The authentic formations, of course, led to many imitations that were the rare and awkward origin but are now more common and sometimes impressive if fraudsters have worked as a team for long hours. Competitions were held but no hoax could not withstand scrutiny.

To distinguish a genuine pattern of a hoax, investigators mainly focus on bending the foot of the plant, which must always remain uninterrupted, something the imitators are unobtainable because they reduced their design in all directions during manufacture.
Freddy Silva sometimes uses infrared photographs that need to appear signs of disturbances in the spread of water into the ground if the design was actually created by rays inducing heat.

Debbie Benstead and other researchers often identify clearly a metallic taste in their mouth when they walk in authentic training which could be due to residual electromagnetic field interfering with the filling of teeth inside of their mouth. In 1991, according to research by George Wingfield, the CIA and the British Ministry of Defense had secretly persuaded two retired David Chorley and Douglas Bower, declaring that they were the authors of crop circles observed up.

Now, without providing evidence to support their statement.

8 / PLASMA & HOLOGRAPHIC TECHNOLOGY

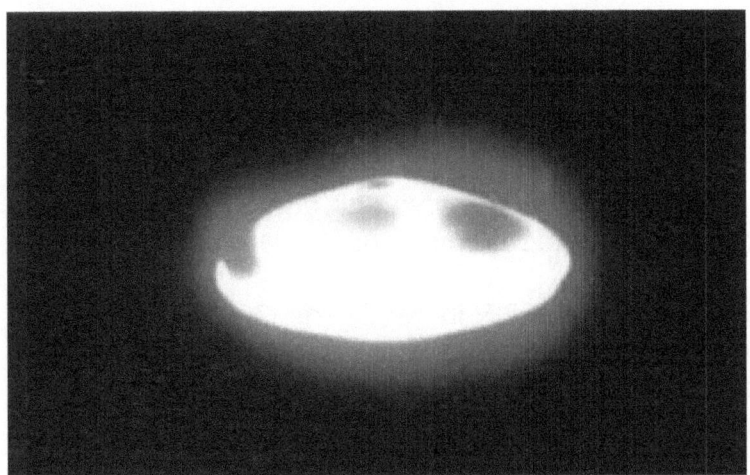

The plasma concept appeared in 1928. A plasma is a fluid consisting of electrically neutral gas molecules, positive ions and negative electrons.
In short, this is an ionized gas emitting photons due to the ionization and therefore more or less bright.
There are three main mechanisms to ionize a gas.
Thermal excitation causes an atom such collisions can give rise to an electron pair + positive ion.

This couple is unstable and tries to recombine. But if the temperature is sufficiently high and sufficiently high density, each recombination quickly followed by a new ionization and the plasma is able to sustain itself.
The temperature required for this process is at least 10,000 ° C (18,000 ° F).
Using a powerful and a laser focusing lens, it is possible to ionize the air locally at the focusing point.
If, for example, the lens has a focal distance of 1 meter, a plasma bubble forms "miraculously" at a distance of 1 meter from the lens and seems to float in the air.

Using an infrared laser, whose rays are normally invisible to the naked eye, the result is spectacular.
But for this project "UFO" wide area, it would use a powerful laser and a lens capable of focusing the projection distance.

It is more efficient to use a converging laser array to a given point in the sky. The first high energy lasers operating at the carbon dioxide (CO_2) and in the infrared scale.
They appeared in the US in 1968. The CO_2 was introduced at one end of the laser, while remaining non-toxic gases were expelled from the other side.
The first attempt of conversion of this weapon in a transportable weapon was made by the US military. Towards the middle of the seventy years, a CO_2 laser with a power of 30 kilowatts was mounted on a tracked vehicle, the LVTP-7, to create a "mobile test unit".
At the end of the seventy years, the German company Diehl designed a similar prototype, the HELEX (High Energy Laser Experimental). It was an armored vehicle of 28 tonnes for carrying a high power CO_2 laser with a power of several megawatts and whose scope would have reached 10 km in clear weather.
The consumption of CO_2 required to allow 50 to each output laser shots.
The US military has continued with new tests of a "close combat weapon laser" or "Roadrunner", a vehicle designed to destroy the sensors and night vision equipment of the enemy.

Then came the "Airborne Laser Laboratory," a Boeing aircraft carrying a 400 kilowatts of laser that succeeds in 1983 to destroy several air to air missiles "Sidewinder". Regarding the use of such a weapon on board a ship, the humidity problem emerged, which could significantly

disrupt the projection of the laser beam. In France, it was not until 1986 that the DGA (General Delegation for Armament) launched the project Latex (Laser associated with an experimental turret), using a laser of 10 megawatts.
If all these devices were (or still are) simple prototypes, they could nevertheless be responsible for observing several UFOs.

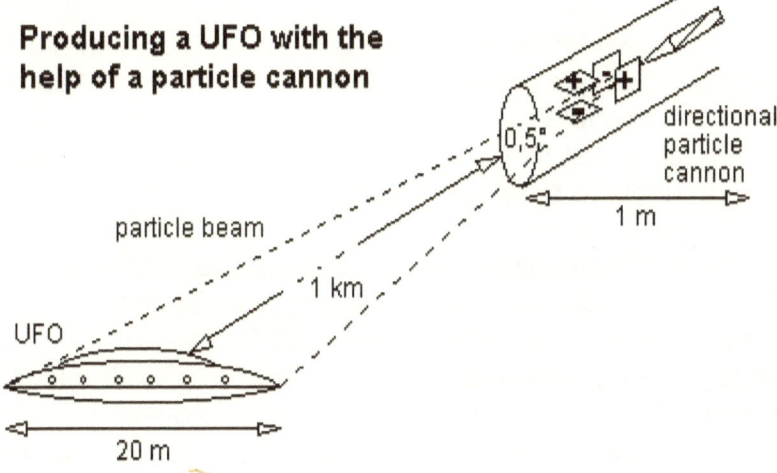

Let us remember that the discovery of the laser only dates from 1958 and it is only from that date that could have been used deliberately to produce fake UFOs. This remote plasma production technique is not old enough to have been used since 1942 in Los Angeles, when the first historically documented event of an unidentified luminous phenomenon simulating an air attack on a clear day. certainly not a case of projection on the background of clouds. electric ionization: this occurs when a strong electric field is applied to a gas.

The electrons liberated by the electrostatic forces are then accelerated and acquire a large kinetic energy,

which allows them, when they collide with other atoms, to spread the ionization process.
Lightning is a good example of creating this type of plasma.
Ionization radiation: this occurs when the atoms are subjected to electromagnetic radiation whose photons have a higher energy to the ionization threshold.
This occurs naturally in the upper atmosphere when ultraviolet photons from the sun ionize the gas atoms of the ionosphere layer.
Since 1991, we knew that the scientists of the Strategic Defense Initiative President Reagan understood in 1981 that it was possible stimulate fluorescence of sodium layer 90 km in height by means of a laser beam (a photon beam) in order to create a bright spot.

This technique of producing an "artificial star" (also a UFO ...) was rediscovered in 1985 by two French astronomers and has since been used for focusing telescopes. The beam can also be used to scale the high frequency (radio waves) or microwave (microwave). Focusing these waves can be obtained at a specific point in space from an array of antennas transmitting phased waves. Thanks to the technique of "synthetic aperture", this matrix can simulate the effect of a giant lens with a long focal length.
During his acceptance speech for the Nobel Prize, Pyotr Kapitsa described in 1978 the Soviet experiences remote plasma generation with powerful microwave. The US Air Force uses this technique to produce "atmospheric ionospheric mirrors" (AIM) that allows them to bounce radar waves so as to explore beyond the horizon or do the same with radio waves, allowing they communicate between two specific positions. These mirrors also allow them to intercept or jam enemy communications.

Everyone can experiment by itself creating a plasma using a microwave beam emitted by a magnetron.
To do this, simply place a fresh grapes on a saucer in a microwave oven, the grapes cut in half but the two halves being always connected.
Soon, grapes and ignites the flames series thus created which are just balls of plasma - rises to the top of the furnace where they survive for some time through the stimulation of microwaves whose frequency is 2 here , 45 GHz (gigahertz).
Microwaves have been artificially produced by Heinrich Hertz in 1887, the magnetron was invented in 1921 and the klystron in 1938.

As the first "maser," the equivalent of a laser to the microwave, it is appeared for the first time in 1953. This technology is probably still in its infancy, was thus already available in 1942. To generate a plasma, photon beam can be replaced by the emission of other particles such as protons or electrons.
A synchrotron can generate enough energy proton beam to traverse a distance in the atmosphere by releasing only very low radiation, caused by a slight loss of energy.
When this energy is below a threshold because of these losses, the protons can no longer move in the atmosphere and the remaining energy, always important, then ionizes oxygen and nitrogen to form a bright plasma ball: point in the sky.
By adjusting the energy of the proton, one can decrease or increase the distance at which the luminous plasma is formed.

Rapid adjustment back and forth can thus give the illusion of a streak of light in the sky.
Similarly, by changing the amount of emitted protons, can decrease or increase the plasma light intensity.

Finally, one can play with the direction of fire in order to produce a specific light form by applying a sweeping motion.

This type of production is within reach of the army, which is capable of generating light phenomena either the ground or an aerial platform. A calculation of Tom Mahood we found on its website tells us that a synchrotron medium capable of generating a continuous beam of protons with an energy of 500 MeV (mega electron volts) would be able to produce a luminous plasma at a distance of 1200 meters.

This beam would lose 3 KeV (kiloelectronvolts) for each centimeter traveled before releasing 100 KeV per centimeter during shutdown.

The light intensity of the beam per centimeter would be equal to 3% of that of the plasma ball.

The latter would have ten meters in diameter, or 1% of the distance in our example.

These calculations were made using the Bethe.
It seems however that there must be a phenomenon left aside by this formula, so that the energy can be reduced by a factor of 100, thus effectively limiting the size and weight of the synchrotron to use. It turns out that the first particles emitted heat the air passing
through them, causing expansion of the air before stopping, which allows the particles that follow to travel farther since they meet less resistance.

In this way, a kind of low density tunnel is excavated in a split second in the atmosphere to the farthest point possible, where the UFO is thus produced and can be maintained with much less expenditure of energy.

One might object that the particles can be accelerated as a high vacuum, which raises the question of how they are projected into the atmosphere.

This problem can be overcome by using a material permeable to protons at the point where the beam out of the synchrotron.
Nickel, tantalum or Kapton, for example, are able to perform this task.
However, they must be cooled because the passage of particles causes a sharp increase in temperature.
Tom Mahood said he submitted his hypothesis us several physicists working in particle physics and that they saw no objection to that.

It is possible that using electrons instead of protons can produce the same results while consuming less energy. However, because the electron has a mass of about 2000 times less than that of the proton, it will certainly have more difficulty penetrating deep into the atmosphere before being shaken by a kind of collision.
The US military currently exploring actively the concept of "charged particle beam" (CPB) composed of ions or electrons able to move in the atmosphere at a speed close to that of light, as well as that of "beam neutral particles "(NPB), composed of hydrogen atoms and deuterium, which can be used in space to combat against

ballistic missile through the SDI.

It will certainly be more difficult to penetrate deep into the atmosphere before being discussed abruptly by any collision. The basic principles of particle gun used could be similar to the operation of the electron gun used in our TVs. A particle beam with a horizontal and vertical scanning is used to draw a crude form over long distances.

The shape can be moved as a whole and can simulate an erratic flight or have amazing speed laps if the particle gun is controlled by a motor.
This motor, directed by computer, can be connected to a radar system locked on the target (control, vehicle, airplane) in order to follow it automatically.
At distances greater than a few kilometers (it is assumed), the form is somewhat limited to light points or drops, due to lack of sufficient focusing capability. Over the years, technology has evolved, the shapes have been refined and now, instead of fixed screenings,
animated projections became possible.

Recall that if there is an array of antennas for transmitting

radio waves or microwaves, the plasma thus produced can be moved as a whole by electronically controlling the phase or each antenna transmission frequency .
The first type of accelerator of high energy particles, called cyclotron, appeared in the early 30s in the United States.
The energy that can be transmitted to protons was so intrinsically limited to 25 MeV. However, it was possible to consider sending heavier ions that protons and therefore more energetic at the same speed transmission, such as hydrogen isotopes (deuterium) or helium (3He, 4He), made more heavy by the presence of neutrons in their nucleus.
This technology was thus available in 1942 despite some reservations about the limited energy of the particles emitted and the weight and bulk of the required cyclotron. A few years later, the Synchro, an improved version of this machine capable of transmitting particles to an energy of 000 MeV was 1 unveiled in 1945, still in the United States. Today, the largest synchrotrons achieve an energy of GeV 1000 (gigaelectronvolts).

These are the three basic mechanisms to produce a long-range luminous plasma.

It will be objected that night seen UFOs sometimes seem opaque or even metallic.
This opacity printing could be obtained by a cannon performing scanning with just bright enough plasma to simulate a metallic gray color.
Albert Budden stresses in this respect that the light passing through a humid atmosphere subjected to an electromagnetic field may give the appearance of a metal surface, since the refractive index of a material, in this case droplets water suspension, changes in the presence of an electromagnetic field.

Why develop such equipment?

We can list several possible uses: Produce atmospheric ionospheric mirrors.
Produce radar decoys or visual decoys to deceive the enemy to illuminate enemy a site for a long time, as if it were day.
Mark an enemy target in order to guide a missile or turn an enemy missile towards a false target and explode.
Delete the toxicity of a poison gas spread by the enemy, causing a reaction with the plasma produced.
To disrupt or destroy long range electronic equipment, electric or electromechanical (motors) beam particles.

To cause fires or cut electric cables by melting ... Blinding, burn or kill an enemy soldier. However, several questions remain concerning the firing of a luminous plasma. We indicated in italics some possible response tracks:
Which weight and volume gun are necessary, depending on the intensity of the observed phenomenon, its size and the distance between the firing point?
To illustrate this issue include the example of "experiments beams aboard a rocket" (BEAR), successfully conducted in New Mexico in July 1989 as part of the Strategic Defense Initiative.

The linear particle accelerator installed in the rocket was placed in a tube of 4.36 meters long with a diameter of 1.12 meter.
It seems that the particles have been issued with an energy of about 4 MeV.
The weight of a particle accelerator is generally greater than 500 kilograms per linear meter.

She produces the form an electromagnetic field?

The local concentrations of positive and negative electrical charges in plasma create electric fields and induced magnetic fields.

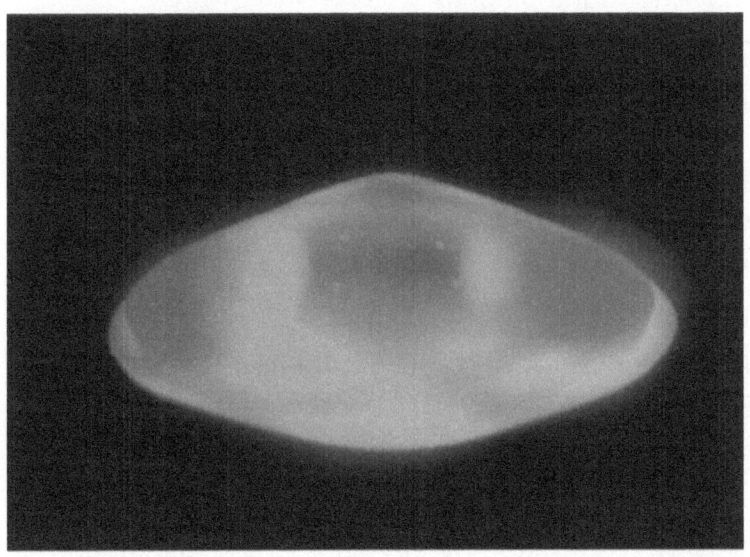

Can the shape emit X-rays may radiate witnesses?

Hot plasmas can emit dangerous X-rays to the witnesses who are nearby.

The shape emit dangerous ultraviolet rays?

The sun is a typical example of a hot plasma ball emits ultraviolet rays that can lead to cancer.
On a smaller scale, tanning lamps also produce ultraviolet radiation from ionized gas (plasma) in a glass tube.

The shape emit microwave?

It is most likely that the emitted light rays overlap towards longer wavelengths, including infrared and microwave.

The shape she emerges from her?

It so happens that a plasma emits a hiss or hum.
This is called "plasma waves".

Can the form give a breath of air?

The ionization of the air and collision cascade molecules sometimes generate an electrical wind having the strength of a light breeze.

Can the plasma ball produce an odor, for example that of sulfur (which is traditionally associated with the smell appearances evil)?

It is sometimes accompanied by a strong, unpleasant odor characteristic of ozone or nitrogen oxides.
The microwaves emitted by the plasma can also cause oxidation of the sulfur into the atmosphere.

Can the plasma ball burn on contact (vegetation, witnesses...)?

Plasma is a gas heated to hundreds, thousands or millions of degrees. It is therefore normal that it burns on contact, even at a distance, depending on the temperature.

Is it still possible to reach some type of plasma with the hand without getting burned?

A plasma produced by a high energy electron beam can be maintained "at a temperature near room temperature."

Indeed, although the electron temperature can reach 700 ° C because of the very rapid movement of electrons, the low thermal motion of ions can impart a temperature below 30 ° C to the whole plasma.

By day, can generate the plasma ball a shadow? Can it also be the night when situated between the moon and the witness?

Depending on the plasma type, a portion of the incident light will be reflected, a portion is absorbed and part will be transmitted.
If the light is mostly reflected or absorbed, a witness may see a shadow.

Can the plasma ball to be illuminated by the headlights of a car?

Yes, for certain types of highly reflective plasmas.

Can the shape generated be detected by radar?

An ionized plasma reflects long wave (radio), but it can easily be traversed by shortwave (TV, radar) if the density of electrons is insufficient.
Plasmas higher density produced by an electron beam used to reflect a frequency of radar wave below 10 GHz and therefore can be used as a "mirror" high speed radar. In the atmosphere, the "artificial ionospheric mirror" may reflect frequencies up to 2 GHz, a ratio of the US Air Force.

Can the shooting through the clouds and how the shape generated if she would include in the rain ?

A beam of particles such as protons can pass through

clouds. Microwaves also pass through clouds with the exception of certain frequencies.

The cloud layer does not significantly it reduces the possible distance for the shooting?

The laser light beams or emitted in the bands near infrared or ultraviolet, they can not of course go (or are heavily distorted).

Can the plasma shot pass through a window, even a section to create a form of light in a room?

It seems that a beam of particles such as protons can not pass through a window or a door.
Microwaves can pass through a window or a flap, provided it is not metal. A laser beam can of course pass through glass but not through a shutter. Finally, a ball of plasma generated outside can pass through glass, as seems to happen sometimes in the case of ball lightning.

Is it atmospheric constraints such that the presence of dust or pollution, humidity or wind?

Dirt could certainly prevent particles from reaching their destination.
The interaction of dust and the particles could also make the beam more apparent. The dust does not disturb the microwave, while a laser beam would be seriously disrupted. If plasma is generated by a particle beam, we have seen that this beam should be somewhat light.
If it is produced by microwaves or infrared laser is invisible to the naked eye, except perhaps in the case of exceptional weather conditions.
Because of his previous appearance, since its operational availability from 1942, particle gun technology which we

prefer, and on which we intend to focus in the remainder of this study.
Thus, we will regularly reference to a "particle gun" whenever we discuss the artificial generation of a plasma ball in the air.

8.1 / HOLOGRAPHIC PROJECTOR

(54) Title: HOLOGRAPHIC PROJECTOR

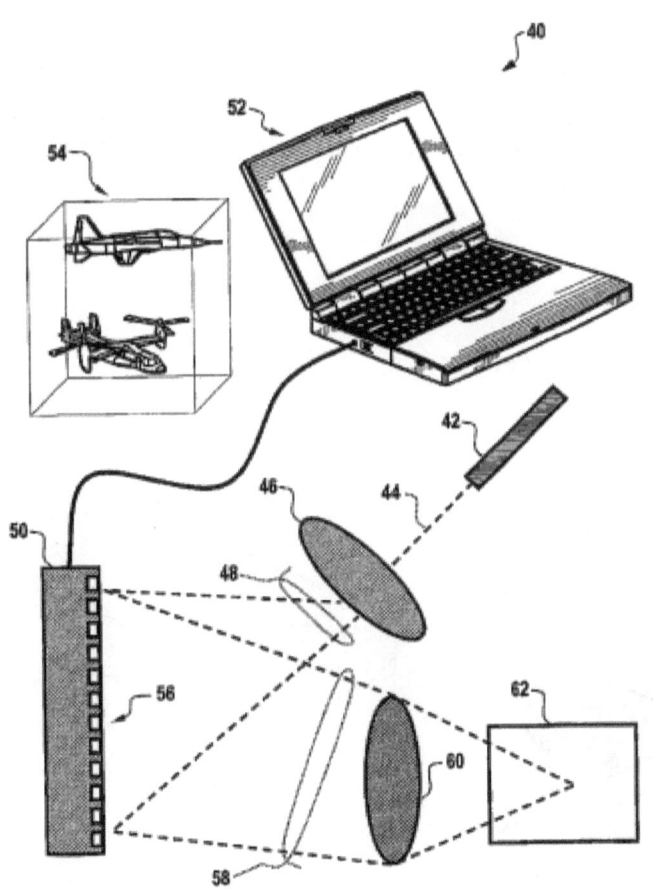

The US military has explored the idea of using holograms during the 1991 Gulf War to deceive the Iraqis, but has not pursued for technical reasons. One idea was to project a hologram of God (Allah) several hundred meters above Baghdad, but that would require a mirror in space of more than one square kilometer, as well as huge projectors and sources energy. Moreover, the representation of Allah is strictly forbidden by Islam, it would have made the event not possible. It is quite likely that this project have served for the operation of 11/09/2001 ...
But that's another story.
The first patent on this technology was introduced in 1932, and that of the " Sky Projector " in 1973. Finally a patent on a projection method in the atmosphere is published in 2004.

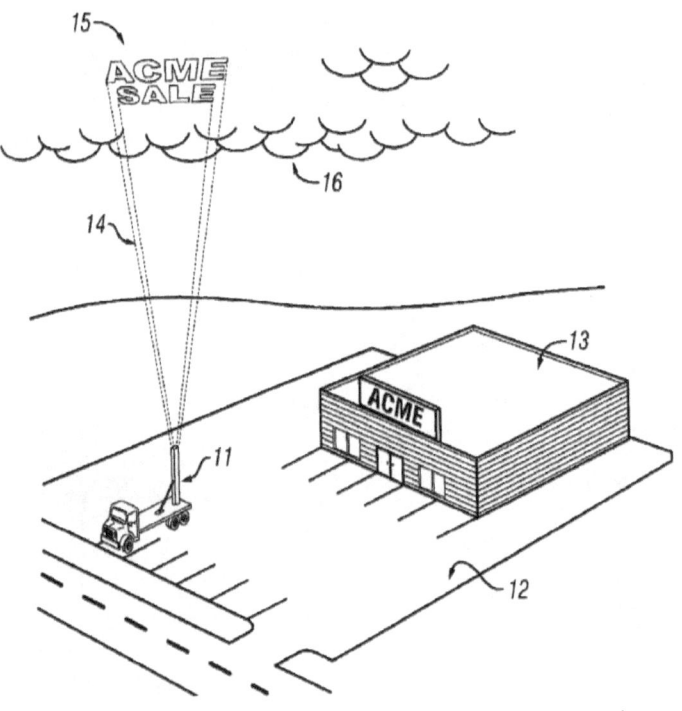

The holographic projector displays a three-dimensional visual image at a desired location, removed from the display generator. It can be used for psychological operations and perception management stratégique.Il is also useful for optical deception and camouflage, providing a momentary distraction when engaging an unsophisticated adversary.

The capacities

•Projection accurate 3D visual images in a selected area
•Camouflage against the optical sensors

Enabling Technologies (MCTL)

•4.1.4 Image processing (holographic views)
•10.1, lasers
•10.2, Optical
•10.3 Power Systems

white papers
•Q, and special humanitarian operations
•N, strategic attack

9 / STRANGE DEATH

Below is a non-exhaustive list of persons whose death associated with UFOs, may seem more or less suspicious.

Captain William Davidson and Lt. Frank Brown

• In late July 1947, after having investigated the case Maury Island, Lieutenant Frank Brown and Captain William Davidson of the intelligence service of the USAF perished returning from their mission, following the fall of their aircraft whose left engine caught fire.
The plane, a B-25, however, was known for its security and was suspected sabotage. Kenneth Arnold (the observance of 24 June 1947 to the origin of the term "flying saucer") returning from the same survey aboard his personal aircraft had, meanwhile, landed in disaster when his engine stalled abruptly. He had to refuel and realized afterwards that the fuel inlet valve was closed at that time.

As for him, he was well in his sabotage.

Morris Jessup

• The Morris Jessup astronomer committed suicide April 20, 1959 by connecting a pipe between the muffler his van and a vehicle window.
It was published in 1955 "The Case for the UFO" after which a mysterious Carlos Allende began a correspondence with him made of complex and obscure revelations abused his credulity and led him to depression. C.Allende later proved to be a pathological liar behind the legend of the "Philadelphia Experiment," which claimed that the USS Eldridge ship had suffered in 1943 a magnetic teleportation.
This story has been known for a resounding success with a naive public always more numerous. Morris Jessup seems to have been the victim of a simple scam.

Edward Ruppelt

• Captain Edward Ruppelt died in 1960 of a heart attack at the age of 37 years. He tried to direct in 1952 the Commission of Inquiry "Blue Book" of the US Air Force with great intellectual honesty, but had resigned discouraged in 1953. His book "The postponement is unidentified flying objects" published for the first in 1956 was reissued in 1959 after being watered down and shifted to a skeptical position, perhaps as a result of pressures he was subjected to the aircraft manufacturer where he worked and whose army was a very big customer.

Waveney Girvan

• When he became editor of the Flying Saucer Review, Waveney Girvan transformed it into a serious magazine that ceased thus ridiculing the UFO phenomenon and promote the most absurd stories as fabrications of George Adamski, aims to which she had
yet been created. secret man careful, Girvan could confided to no one and left no record at
the newspaper's office. He died prematurely in days of a galloping cancer October 22, 1964 and all the records he had in him disappeared.

Frank Edwards

• Here is what was written in 1969 French ufologist Aimé Michel about the American journalist Frank Edwards. "In his book" Flying Saucers, Serious Business "(Robert Laffont 1966) F.Edwards recounts in detail the persecution he was subjected when he
began to exhibit, document in hand, the activity of the CIA in the field of UFOs, how they threatened to shut him up, how he lost his job ...

But, he adds, in essence, it is too late to shut the mouths now that I got to certainty and I gained my financial independence, and I defy the authorities to silence me. The last pages of his book are prophetic "on realization comes, the time is closer than we think" he writes textually.
He thought about the need for the secret services finally put on the table. He hoped to force them. The "outcome" came another way: Frank Edwards published these lines,

and soon died of a "heart attack" in 1967 ".

James McDonald

• James McDonald, Dean of the Institute of Atmospheric Physics of the University of Arizona, professor of meteorology and ufologist recognized, committed suicide in 1971 after failing the first time. From 1966 he set the goal of interest to the scientific community to the phenomenon "
UFO " and there he displayed considerable energy. Meanwhile, he was already struggling for the preservation of the ozone layer and also campaigned against the Vietnam War. He had uncovered
some disinformation maneuvers of the US military regarding UFOs and some defamatory texts tried to tarnish his memory. He was, apparently driven to suicide by dishonorable accusations, ironically on the fact that he was interested in UFOs.

René Hardy

• Dr. René Hardy, founder in 1963 of the GEPA (Grouping Aerial Phenomena Aerospace Study and unusual), died "suicide it seems" a bullet in the head in June 1972, "a week after he said" I discovered the defect in the armor of UFOs, it's fantastic ... I will speak next week at home. " For him, there was no next week! ".

James and Coral Lorenzen

• James and Coral Lorenzen, founders of APRO (Aerial Phenomena Research Organization), which has 3,000 members, died of cancer in 1986, respectively, and respiratory problem 1988, at an age close to 65 years. Coral had written " The great flying saucer hoax " in 1962.

Scott Rogo

• Scott Rogo, author of "The haunted Universe," "UFO abductions" and co-author Ann Druffel of "The Tujunga Canyon Contacts" stabbed died at his home in Northbridge, near Los Angeles, around the 15th of August 1990 at the age 40.

Karla Turner

• "The remarkable work of Karla Turner's investigation was stopped prematurely by his death on January 9 1996, at the age of 49 years. ". She was investigating alien abductions in which the military where members of the secret services were clearly involved. She was struck down by a particularly virulent form of breast cancer immediately after she herself had a RR4.

Colonel Uyrange

• Colonel Bolivar Uyrange, retired Army Brazilian Air, was found "suicide" at his home in 1997 after confirmed in July that his team had photographed and filmed UFOs of the deadly wave of Corales, in the end 70s he appeared on television and was preparing a lecture tour in the country.

James Forrestal

James Forrestal was found dead the morning of 22 May, on a roof located thirteen floors below his room. He was connected to
Majestic record 12. It is assumed that he had a desire to talk about secret plans.

10 / MISSILE DISABLED

Retired military officers have made public their testimony about the repeated incursions of UFOs on US military nuclear sites.
These UFOs have disabled several times nuclear missiles. It was September 27, 2010 that was held the National Press Club conference where the military has been able to speak about it.
Broadcast live on CNN, the video lasts 1:27, is subtitled in French since October 13 with EP & ProjectAvalon.net for JLG. A site pro UFO. We now know that Boeing has developed a drone that can disable everything in its path.

Computers, servers, appliances or lighting, everything goes.

Named Counter-electronics High-Powered Microwave Advanced Missile Project (CHAMP), it was tested in Utah, the United States successfully. This is a harmless weapon in humans and may include several buildings individually. If officially it is only a prototype, it is not said that such missile is actually made available to some countries to offset some strategic areas.

When we know that the technology has the public today is still available decades before by the secret governments, it is easy to understand that disabled the missiles probably have to be in this kind of gear.

11 / WAS THE FLYING SAUCER INVENTED BY NAZI?

If some projects are now proven Nazis, others, often under science fiction, however, are described in great detail in some books and digitized on the Internet.
Among the Nazi secret weapons most famous are, in particular V1 and V2, real ballistic missiles of the time, these unmanned aircraft were taking explosive charges of several hundred kilograms to distances up to hundreds of kilometers.
Less is known about the other Wunderwaffe the "Formidable Weapons," developed in secret by the Nazis and who had to win them the war. In the aeronautical field in particular they abound.
Some remained at the prototype stage, others would become models mass produced secretly recovered by the Allies destroyed by the SS before the end of the war, or mysteriously disappeared.

View a loan V2 to be installed on its launch pad. The Germans fired a few V2 London, from occupied France in 1944, which did a lot of physical damage and psychological.

Among them include e project Arado Ar E-555, a flying wing jet capable of carrying a hypothetical atomic bomb and with a sufficient range to hit the East Coast of the United. First born of Strategic Bombers Contemporary reaction, its developers were recruited by US intelligence through Operation Paperclip.

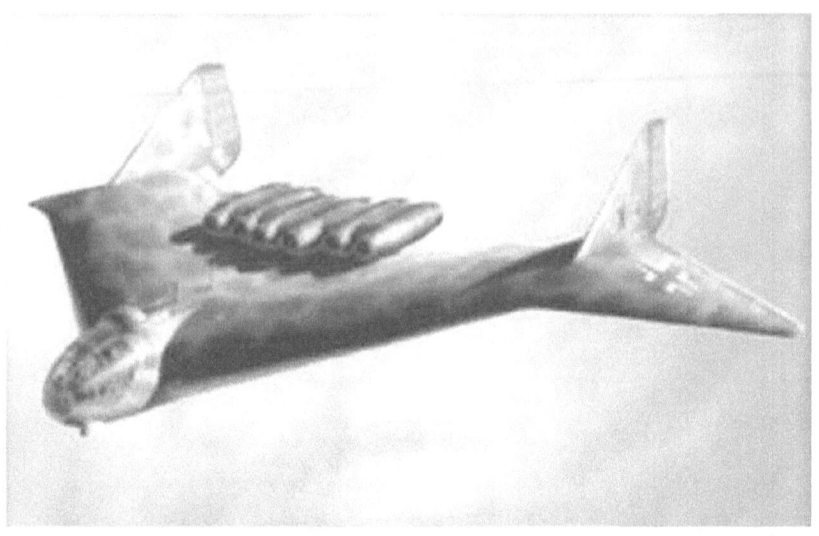

The silhouette of the Arado Ar-555 E is not reminiscent of the famous American stealth bomber B-2. Other weapons, other flying machines, have also been developed in great secrecy.

11.1 / ESOTERISM NAZI

The fact that Adolf Hitler chose the swastika (swastika) as a symbol for the NSDAP (National Sozialistische Deutsche Arbeiter Partei - German National Socialist Workers Party, the Nazi Party)
is a strong indication of his interest in the esoteric, mysticism and the occult.
Recall that the Swastika is an ancient symbol that is found in many pre-Christian cultures (Hindu Indo European, and even in North America in some pre-Colombian civilizations).

This lithograph of the Thule Society shows the use of certain symbols in 1919. The Nazis recover them widely.

The Vril company

Several authors have argued that the Vril-Gesellschaft (The Coming Race) was a secret community of occultists in pre-Nazi Berlin. The Berlin Vril Society was in fact a kind of inner circle of the society of the Order of Thule. It was also thought in close contact with the English group known as the Hermetic Order of the Golden Dawn. Legendary and mythological for some real good for others, the Vril is a form of energy possessed by this extremely powerful underground race that are Hyper-Boreans who live in Ultima Thule.

It has, for the first time, was exposed in The Coming Race (Race to come), novel written by Edward George Earle Bulwer Lytton in 1870.
There is only one primary source of information about the company Vril: Willy Ley, a German engineer who fled the United States in 1933. In 1947 Ley published an article entitled " Pseudoscience in Naziland " (Pseudoscience in Nazi countries). After a description of the aryosophie, Ley writes: "The group (...) was literally founded on a novel [of Lytton].

This group was called Wahrheitsgesellschaft - Society for Truth - and probably met in Berlin to devote to research Vril. A Johannes Täufer published in 1930 two trials Vril, designated as a cosmic primordial force and taking the same frame as the novel of Lytton.
He belonged to a company "discreet" The coming Germany.

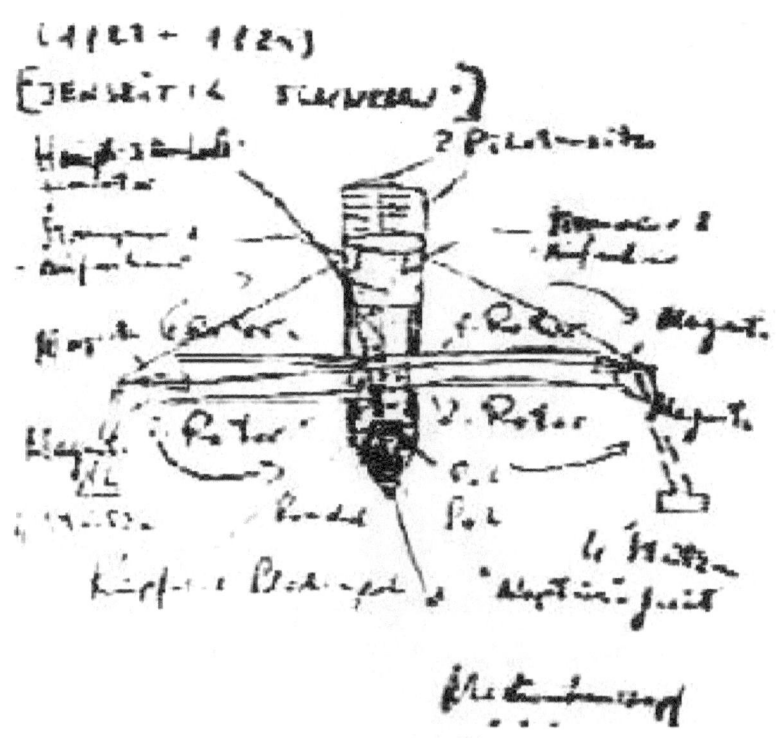

Annotated diagram of what appears to be an attempt to reproduce the artifact, or at least its effects in applied physics.

Antarctica

Expeditions were launched in Antarctica, in the area known as New Swabia. Officially it was for Germany to ensure access for its vessels from fishing in this area rich in fish, unofficially, the Nazis were looking for Hyperborated, some discoveries are in the region Maud land Queen. On 17 December 1938, the Schwabenland ship from the port of Hamburg with 33 people on board. The boat docks in January 1939 (4 ° 15'W, 69 ° 10'S) and field reconnaissance begins. The following weeks, 2 hydroplanes Dornier Do J of the ship, the Passat and the Boreas perform a dozen flights, crisscrossing the area and conducting more than 11,000 aerial photographs. A temporary database is installed and three Nazi flags are planted.

The expedition members plant the flag Nazi Antarctica

Can be located inside the poles (it would not make that account is in is due to gravity), this mysterious land is for years a fantasy Nazi secret societies. Indeed theses impossible Hollow Earth, and the possible Concave Earth (see my website about it), tend to show the existence of an underground world full of Agartha the Nazis have succeeded in discovering.

The arguments that Antarctica, the North Pole, Tibet, Peru and Mount Shasta in California are the entrance to a subterranean realm known as Agartha, all have lawyers to defend them.

Companies Thule and Vril and resumed his theses to their account and launched the Ahnenerbe looking for an entry point in Antarctica.

Agharta, the mysterious underground world, sometimes compared to Atlantis, the Nazis have discovered in the bowels Antarctica.

So on the trail of an alleged lost civilization, and in possession of the information to reproduce the energy of the Vril, the Nazis are embarking on a new project: Die Glocke Vril Projekt designed to provide a source of energy and some gear an anti gravitational capacity, ie give them the ability to hover in the line of Wunderwaffen projects we have listed at the beginning. This project has been exposed by the Polish journalist Igor Witkowski in his book " Prawda O Wunderwaffe " released in 2000 which is based on the SS secret documents found in the archives of the Soviet Union.

There is no photograph or plan of the German bell Glocke Projekt.
However, the consensus is to think that she should look at the prototype stage, this facility above.
According to Igor Witkowski, Die Glocke was indeed an anti-gravity system building experience. The site would be located near the coal mine Ludwigsdorf (Ludwikowice Kłodzkie Today, Poland exact coordinates: 50 ° 37 '42.02

"N - 16 ° 29' 39.32" E.).

According to Witkowski, the remains of the experience would still be present on the site. She would have need a very large amount of energy, which is a thermal electric plant has been built near the site in 1941. The bell was deemed extremely dangerous, causing heart disease, mutations and death of animals placed too close her.

The circular installation reinforced concrete could be used to test ring or test zone for "Bell" Nazi.

The Vril become vital for the engine and steering wheel of the new circular devices (Flugscheiben), and the operation of a ray gun prototype energy (Kraftstrahlkanone). The latter never exceeded the stage of the drawing board.

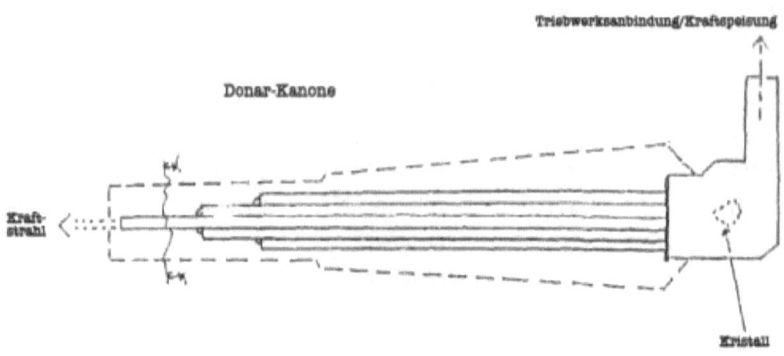

KFK - Kraftstrahlkanone here schematically by the SS engineers.

These technologies, acquired by the Nazis thanks to the huge involvement of their secret societies, their study groups, their scientific and SS, allowed to develop flying machines of a new genre.
By their circular or ovoid, the first to observe, possibly of aviation allied bomber pilots during the war, gave them the name of "Flying Saucers", or flying saucers.

The myth of UFOs was born and he was fine spirits occupy during the second half of the twentieth century.
As early as 1934, an industrial recommends Schauberger to Hitler.
During an interview, he strongly impressed one, who asked that his research on the production of fuel-free energy benefit from all the help. In 1941, Marshal Ernst Udet, commander of the Luftwaffe, had asked him to help solve the energy crisis that knew Germany. In 1943, Himmler ordered to develop a secret weapon system with a team of engineers POWs. This will be the project " Haunebu ".

Viktor Schauberger with a model of his domestic power station (1955).

Viktor Schauberger

The work of Schauberger, would give birth to the Turbine Vortex Schauberger. The turbine components were manufactured by the Koertl factory in Vienna. According to one of its engineers, Aloys Kokaly, he was told that a test model of the turbines had gone through the roof of the factory. It was actually a prototype called RFZ 1. An improved model was launched on 6 May 1945, the day US forces arrived to plant Leonstein (town Grünburg) in Upper Austria where Schauberger lived. Marshal Wilhelm Keitel then allegedly ordered that all prototypes are destroyed. The Soviets reportedly searched his Vienna apartment, carrying documents and plans. A detachment of US Special Forces had seized all the equipment present at his home in Leonstein and would have placed in "protective custody" during 9 months to devise an assessment of its research in the USA. Keitel mysteriously dies 5 days after his return to Europe.

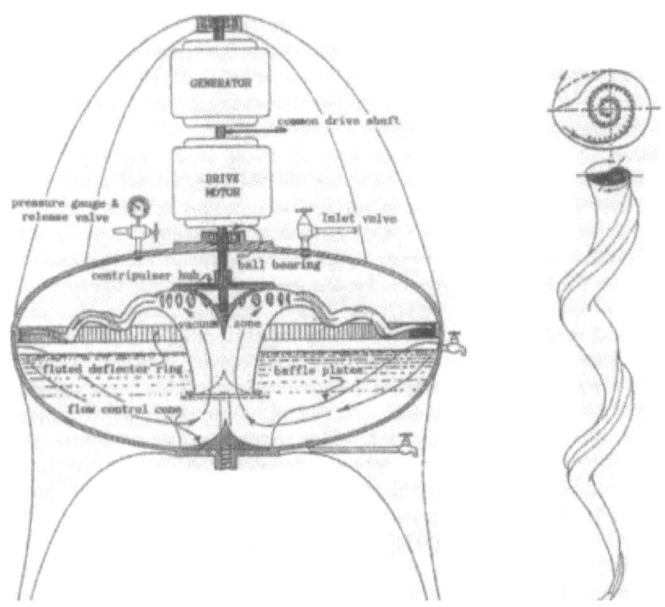

Turbine vortex, called implosion or self-sustaining rotation of Schauberger. This would in fact simply the realization of Glocke Projekt and researching the Vril. Schematic side conducted by US teams after 1945 according to the own work of Schauberger.

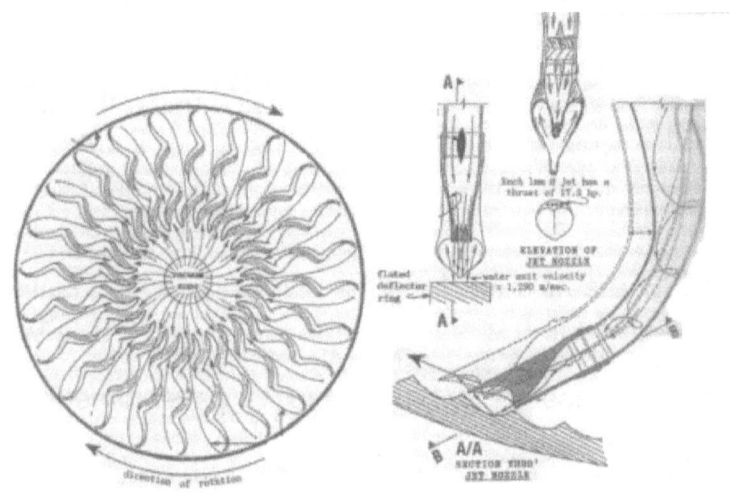

The same Turbine scheme above conducted by US teams after 1945 and after Schauberger own work.

11.2 / WORK ANTIGRAVITY

The Thule and Vril Gesellschaft, through research and Glocke Projekt, are robbed in 1922 an unmanned antigravity demonstrator, called "JenseitsFlug Mashine" (JFM) which means "flying machine on the other side", which will then be dismantled and Messerschmidt stored in plants.
This electromechanical antigravity system is largely based on theoretical work by Nicolas Tesla and the German physicist Levetzow and implosive theories Shauberger. The system comprises two counter-rotating disks (diameter 6 m, 50 m and 7) flanking a third fixed disc (diameter 8 m). The three discs are pierced at their center an orifice (diameter 1 m 80), occupied by a cylinder surmounted by a cone (height 2 m 40),

containing the rotating system and magnetic loading discs.
From 1922 to 1932 the system is simplified and no longer uses a single disk. A controlled device using this drive, RFZ 1 will be tested in 1934 but the trial will prove so disastrous that antigravity electromechanical system will be permanently abandoned the same year.

In 1933, the new Nazi regime places the Vril Gesellschaft and his research under the direct control of the SS Himmler, inside a special section called "Entvloklungsstelle IV" (number four development group) or SS-E-IV.
Dr. Walter O. Schumann, an eminent physicist Technological Institute of the University of Munich and member of the Vril Gesellschaft, Munich creates a first working group under the technical direction of engineer and SS Captain Hans Kohler . For the construction of prototypes, Himmler provides Schumann industrial resources of the aircraft manufacturer Arado in Brandenburg. Dr. Victor Shauberger, physicist and engineer working for the Austrian company Koertl and lecturer at the University of Vienna, joined the Vril Gesellschaft in 1934 after meeting with Hitler.
In June 1934, after the unsuccessful attempt of RFZ prototype 1, a second working group is created in Neustadt near Vienna and under the direction of Shauberger.
Their prototypes are now manufactured by Dornier factories. Team Shauberger includes aeronautical engineers and Shriever Habermohl, engineers specializing in electromagnetic Miethe and Belluzo, the specialist engineer propulsion Andreas Epp reaction and a metallurgist physicist Dr. Erich Wang.
The "Flying Saucers" Nazi therefore the result of years of secret research across the Reich.

A first series, generic, was baptized RFZ (According to sources, would mean RFZ Rund Flugzeuge (round the airplane) or Flugzeuge Reichs (the plane of the Reich). The series of Vril, is also sometimes called Leich Bewapphete Flugscheibe Jager which translates word for word would go something like "hunter Corpses Flyer Sower wheel." the term hunter is interesting because it was indeed at the base to produce a fighter. Relatively speaking, if the Vril are hunters, Haunebu, given their theoretical and operational capacity would be bombers.

RFZ 1 (1934)

The first German saucer thus came into being in June 1934.
It is under the direction of Dr. Walter O. Schumann was born the first experimental circular aircraft on the ground of the aircraft factory Arado in Brandenburg it was the RFZ 1, first with a propeller and d a turbine along with a reactor. Other models RFZ, RFZ 2 to 7 will follow until the end of the war.
During its first flight which was also the last, it rose vertically to a height of about 60 meters and then began to whirl and dance in the air for several minutes. The tail unit Arado 196 which was to guide the device proved completely ineffective. It is with great sorrow that Lothar Waiz pilot managed to land on the ground, to escape and get away from running because the device started to spin like a top before rolling over and be completely to pieces. It was the end of RFZ 1 but the early flying machines PRIL.

RFZ 1 (unconfirmed)

Other RFZ projects were, however, put to the study by the first team of Walter Schumann.

RFZ 2 (1934)

The RFZ 2 was finished before the end of 1934, he had a Vril drive and a drive to magnetic pulses.
Device contours faded when he picked up speed, and the machine lit up in different colors. According to the driving force, it became red, orange, yellow, green, white, blue or purple. He could therefore function and 1941 reserved him a remarkable destiny. It was used as long-range reconnaissance aircraft during the Battle of Britain. It is photographed in late 1941 over the South Atlantic as he headed toward the auxiliary cruiser Atlantis in the waters of Antarctica.
Despite its equipment could not be used as a fighter for the following reason: because of his driving pulse, the RFZ 2 could make changes of direction as 90 °, 45 ° or 22.5 °.

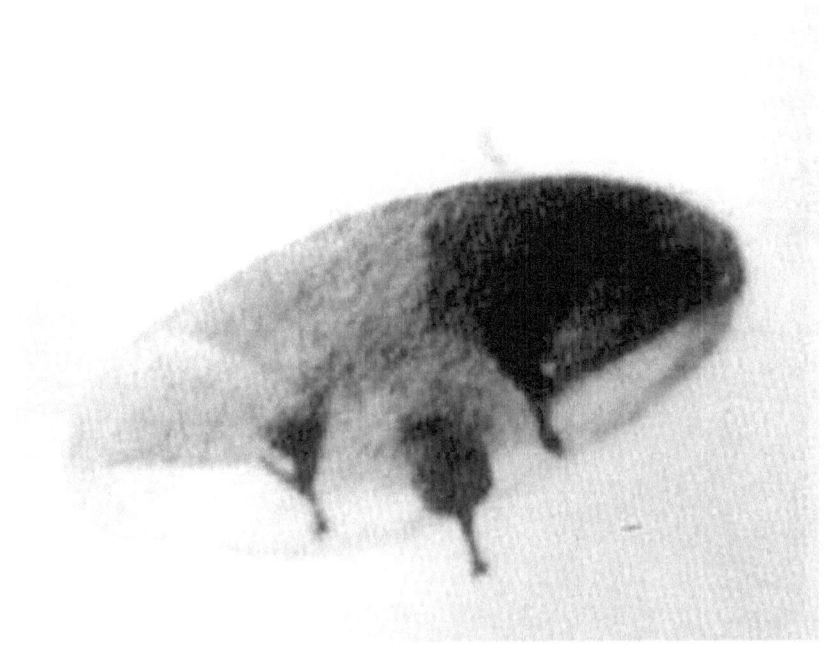

Photo of the RFZ 2 above the South Atlantic

Diameter: 5m
Engine: Schuman-levitators Order: Mag-Yeld-3a boost speed: 6.000km / h (theoretically up 24.000km / h (assumed)
Armament: 3 x Machine Gun MK-108 Shield: Unknown Crew: 1 man
stable flight time: 12 minutes day and night in all weather conditions (assumed) First flight: 1934 Commissioning: 1941

RFZ 3 (1934)

There are only two photos of this machine. Its characteristics are little known, the base where it was developed was probably completely destroyed by the SS.

Snapshot of a RFZ 3, unknown location, probably near the Czech border.

RFZ 4 (1935)

Even more mysterious than the RFZ 3, everything we know about him is that he was the forerunner of the Haunebu 1.

A RFZ 4, probably photographed in test flight above the Atlantic.

RFZ 5 (1939)

The RFZ 5 is only an improvement RFZ 4 and later becamethe Haunebu I.

RFZ 6 (1943)

It was a sort of supersonic aircraft, better known as the V-7 name.
In 1941, Miethe and Shriever begin construction of the first V-7 makes its first hovering near Prague in December 1942 under the name of V-7 FlugKreisel.
At a height of 3m20 and a diameter of 14m40, the apparatus is powered by five turbines Vortex Schauberger, two of which provide the horizontal translation and the other three lead to a rotor with variable pitch allowing the VTOL and sustenance.

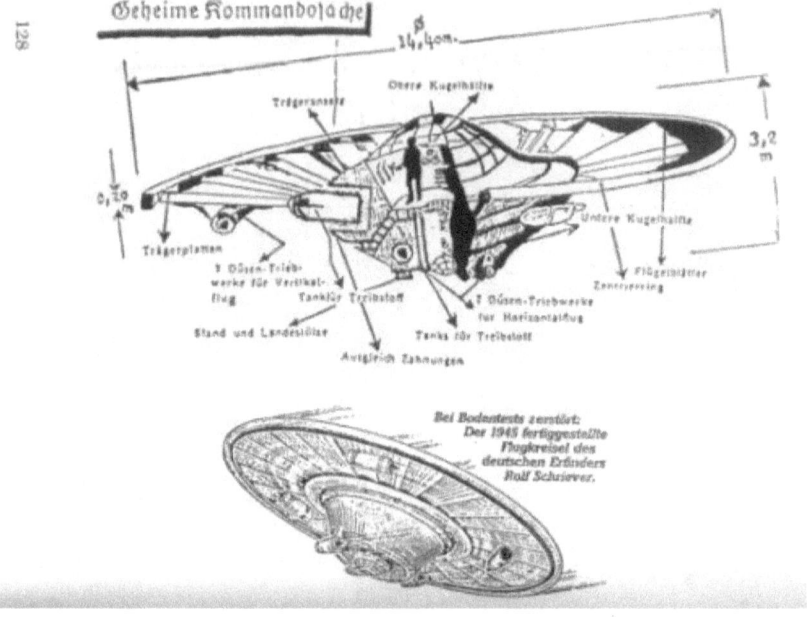

Rfz design Drawings 6 / V-7 FlugKreisel

Diameter: 14.40m
Engine: 5 turbines Schauberger Vortex Control: Unknown
Speed: It never exceeded 100km / h Armament: Unknown Shield: Unknown Crew: 1 man
Term stable flight: Unknown (can be between 5 and 10 minutes)
First flight: 1942

RFZ 7 (1945)

The RFZ 7 is a saucer Coanda effect (Project Omega). This project was under the direction of Andreas Epp and Habermohl engineers.
A version drone with a diameter of 2 m would have flown in 1943.

A single-seater driven by 6 m in diameter was built in 1944 and would have flown to the site of Peenemunde. Version 7 RundFlugzeuge (RFZ 7) or Valkyrie was to have dimensions of 42 m in diameter and weighing over one hundred tons and be powered by BMW 14 reactors to reach 1000 km / h.

VRIL 1

The Vril I, first in the series of "Vril" was experienced from the beginning of 1934 and probably produced until 1942. It was equipped with a Plexiglas cockpit on its top. This is the team Schauberger who was responsible for its design from the engine lévitationnelle implemented by Schumann.

Sketch of Vril I and list of its main features. SS classified document

Diameter: 11.50m
Engine: Schuman-levitators Order: Mag-Yeld-3a boost speed: 2,900 km / h (theoretically up 12.000km / h)
Weapon: Barrels 1 x 80mm KSK on rotating turret; MG 2 x MK-108 Screen: Double Victalen
Crew: 1 man
Stable flight time: 12 minutes day and night in all weather
First flight: 1939
Commissioning: 1944 built in 18 copies

VRIL II

The Vril II is a more powerful version of the Vril I, up to the study in 1936. The cockpit was replaced by a metal pressurized cockpit and topped with a bulb Plexiglas.

Diameter: 10.50m
Engine: Schuman-levitators
Order: Mag-Yeld 3b-drive
Speed: 6.000km / h (theoretically up 24.000km / h)
Weapon: Barrels 1 x 80mm KSK on rotating turret; machine gun 2 x MK-108
Shield: Double Victalen
Crew: 2 men stable flight time: 12 minutes day and night in all weather.
First flight: 1942
Commissioning: 1944

VRIL III

The Vril III is a more powerful version of the Vril II, and equipped with a cannon. Getting to the study probably in 1938.
Diameter: 10.50m (assumed)
Engine: Schuman-levitators (assumed)

Order: Mag-drive-Yeld 3b (assumed) speed: 6.000km / h (theoretically up 24.000km / h) (assumed)
Weapon: Barrels 1 x 80mm KSK on rotating turret; Machine gun 2 x MK-108; barrel 1 x 75mm on rotating turret located in the cockpit. Shield: Double Victalen (assumed)
Crew: 2 men (assumed)
Stable flight time: 12 minutes day and night in all weather conditions (assumed)
First flight: 1943 Commissioning: 1944

VRIL IV

Entered into research firm likely in 1940, the Vril IV comprises a tube above the cockpit which is completely ignores utility. It is an improved variant of the Vril II and III. It's always the team Shauberger which is responsible for its design.

Diameter: 10.50m (assumed)
Engine: Schuman-levitators (assumed)
Control: Mag-drive-Yeld 3b (assumed) speed: 6.000km / h (theoretically up 24.000km / h) (assumed)
Weapon: Barrels 1 x 80mm KSK on rotating turret; MG 2 x MK-108 (assumed)
Shielding: Double Victalen (presumed)
Crew: 2 men (assumed)
Stable flight time: 12 minutes day and night by all time (assumed)
First flight:1943
Commissioning: 1944
Other PRIL were put to the study, often remained the rank of prototypes, some still managed to be successful flight tests with sometimes stunning performance for the time.

VRIL V

The Vril V is probably the most successful team of Shauberger prototype. Aside the study in 1942, its first flights took place in 1944. It reached a speed of 12.000km / h and it even seems that the craft has left the atmosphere several times, allowing himself a few trips into orbit.

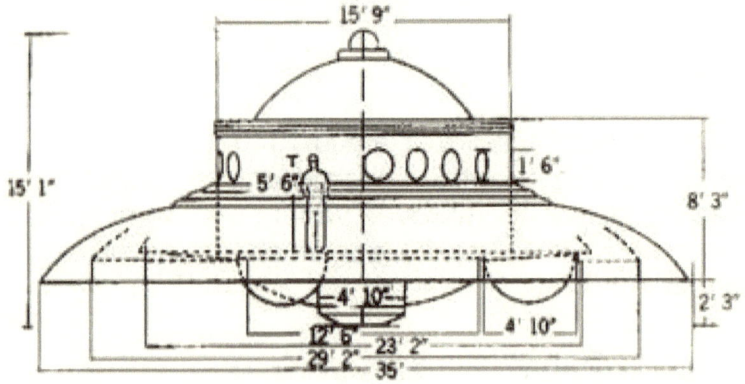

Diameter: 35 meters (assumed)
Motor: Thule Tachyonator (Treibwerk) 7b Order: Mag-drive Yeld-3c (assumed) speed: 12.000km / h (theoretically up 48.000km / h) (assumed)
Weapon: Barrels 1 x 80mm KSK on rotating turret; MG 2 x MK-108; 1 x 75 mm cannon turret located on the cockpit (assumed)
Shield: Victalen triple (assumed)
Crew: 3 men
(assumed) stable flight time: Unknown
First flight: 1944
Commissioning: 1945

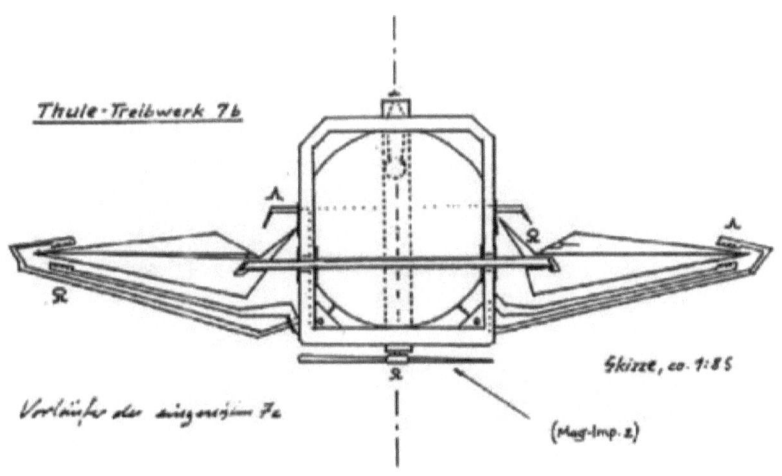

Engine technical study that Thule Tachyonator Vril was to equip gear 5

VRIL VI

The Vril VI will never be built. The project did not exceed the setting in the study. The commissioning was scheduled for 1945.

VRIL VII

In the whole series of Vril, VII is probably the wackiest, but will remain in the state plans. It was a giant ship project in diameter 120m.
Code name "Project Andromeda".
Its commissioning was scheduled for 1946. It would take several Vril and Haunebu saucers and hundreds of crew, a bit like a mother ship.

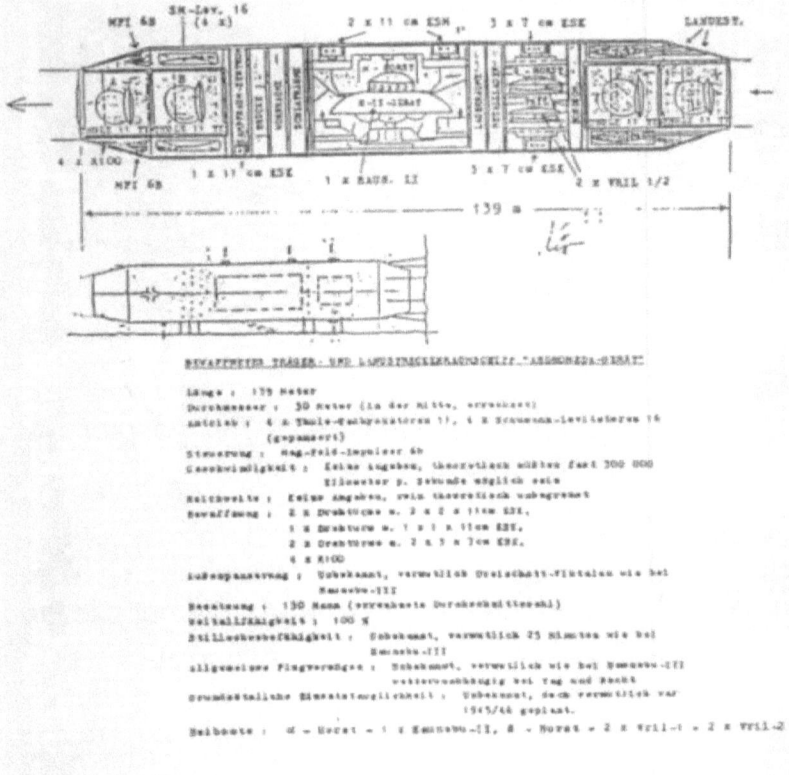

Map of the bunker and Specifications main Vril VII. Document seized by US troops in SS archives.

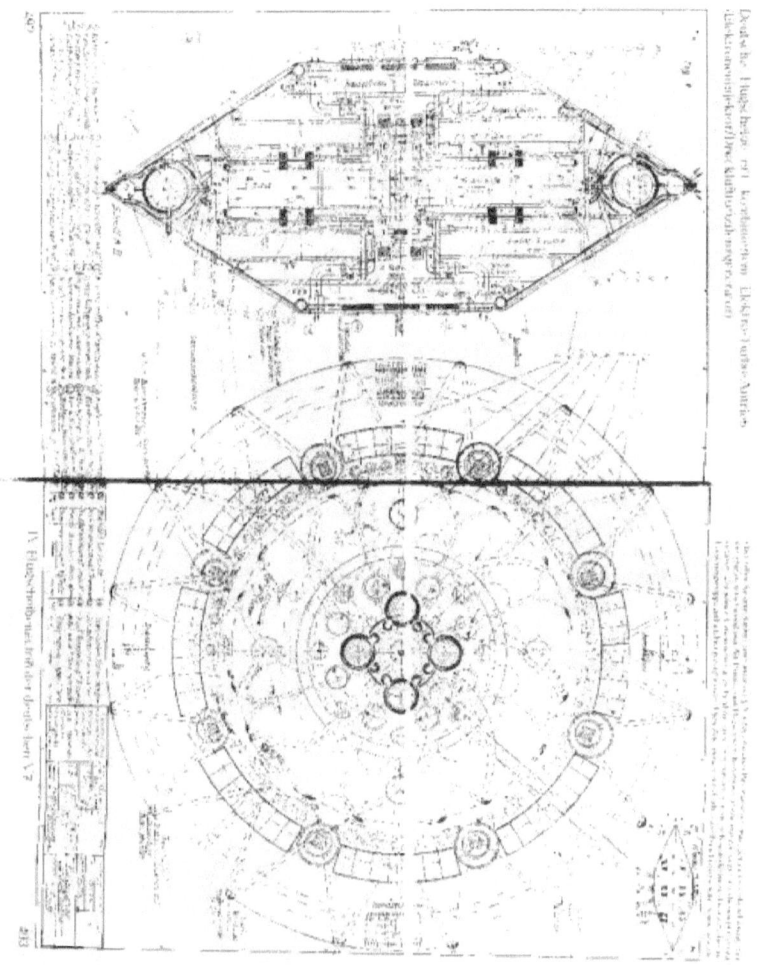

Sketch captioned Vril VII - Project Andromeda.

Some sources, however, indicate that the Vril VII Andromeda was indeed built and it was might smuggle senior Nazi officials as well as some scholar to bases in Antarctica built by successive expeditions to this remote area.

VRIL VIII

The Vril VIII "Odin" will never be built because of the end of the war.
Scheduled for 1946, it seems that its implementation in the study was not conducted. The Armistice of May 8th, 1945 having cut short all the Nazis projects.

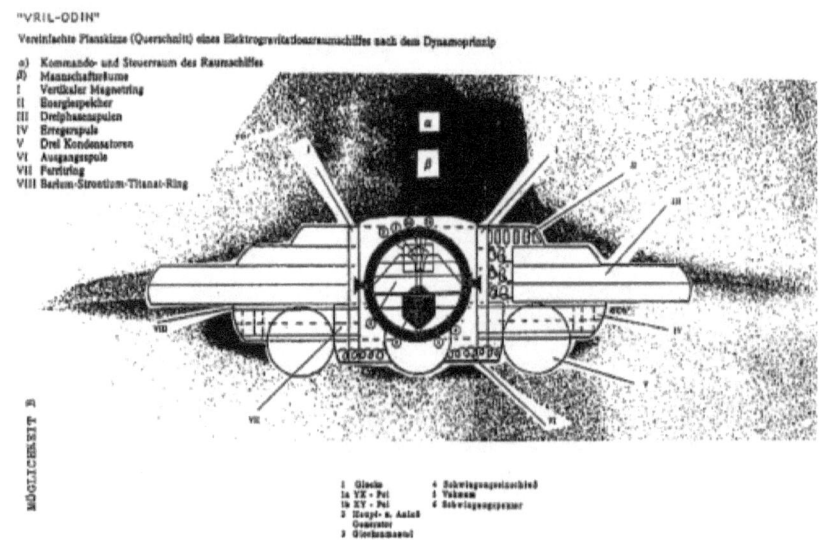

Technical Study Vril VIII "Odin"

VRIL IX

The Vril IX will never be built. It seemed for a role fighter interceptor.

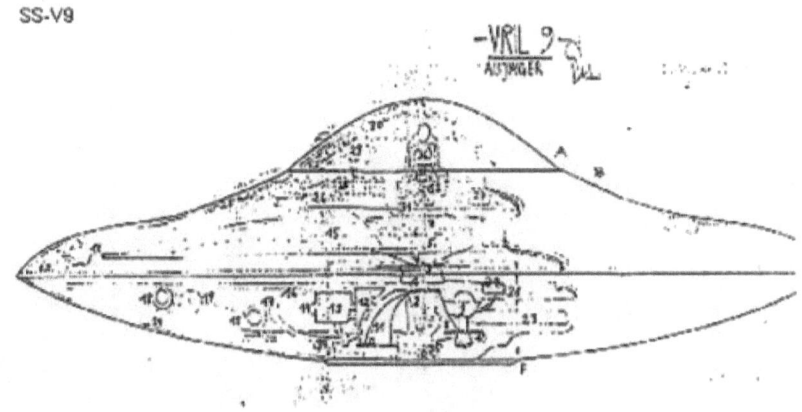

Only a sketch was found.

RFZ 5 - Haunebu I (1939)

The Haunebu I is actually the fifth version of the mysterious series of RFZ. Thule Tachyonator equipped engines, she is unstable in flight and can fly perfectly during 8 minutes. Its armament therefore rendered unstable. The Nazis projected: to use a kind of laser called "The death ray" instead of guns. This "laser" or KFK - Kraftstrahlkanone was theoretically capable of penetrating 100mm of armor, more than any of the allied tanks engaged at the time.

A Haunebu I photographed here in the test flight alongside a fighter Messerschmidt BF109.

Diameter: 24.95m
Motor: Thule Tachyonator (Triebwerk) 7b order: Champs Impulser 4
Speed: 4.800km / h (theoretically up 17.000km / h)
Weapon: Barrels 2 x 80 mm KSK on rotating turret; guns 4 x MK-108
Screen: Double Victalen
Crew: 8 men
stable flight time: 8 minutes
First flight: 1939
Commissioning: 1944

A page from an SS paper which shows the results of flight tests of several saucers models, quantities of gear products and features the main Haunebu I.

Haunebu II (1940)

The Haunebu II was the most famous craft of this series because it was often photographed after the war,

including George Adamski.
During his appearances, some people wrongly believed she was only 10m in diameter. But above all, it is the only device that Americans have recovered whole. It is also the most built version of the series of the Haunebu and some sources say that she was the model for all American saucers built after war.

*Diameter: 26.30m
*Motor: Thule Tachyonator (Triebwerk) 7c
*Order: Champs Impulser 4a
*Speed: 6.000km / h (theoretically up 21.000km / h)
*Armament: 6 x 80mm cannons on KSK rotating turret 3; 1 x 110mm cannon KSK rotating turret
*Shielding: triple Victalen
*Crew: 9 men (20 bit embark person for transport)
*stable flight Duration: 15 minutes
*First flight: 1942
*Commissioning: 1944 (at least 7 copies will be built).

3D modeling of a Haunebu 2

A view of the Haunebu II, the half spheres that constitute its ventral turrets characteristics

Haunebu III (1944)

The Haunebu III was a gigantic version of the series of the Haunebu. She used as an other antigravity propulsion. Its dimensions were absolutely titanic with diameter of 71m.
Its armament was also impressive, combining simple machine guns and electromagnetic guns. Autonomy was also very high, before the materials are completely worn out, the driver could count on 7 to 8 weeks of flight. Although the project was extensively studied, no one knows if it was conducted.

*Diameter: 71m
*Motor: Thule Tachyonator (Triebwerk) 7c and Schumann-levitators.
*Order: Champs Impulser 4a

*Speed: 7.000km / h (theoretically up 40.000km / h)
*Armament: Guns 4 x 110mm KSK 4 on rotating turrets; guns 10 x 80mm KSK on rotating turret; machine guns 6 x MK-108; guns 8 x 50mm KSK
*Shielding: triple Victalen
*Crew: 32 men (can carry 70 people for transport)
*stable flight time: 25 minutes
*First flight: 1945

Haunebu IV (1945)

Much larger than the Haunebu III, she was supposed to be reserved for transporting troops and equipment. All that remains of this project and one single draft. It seems further that this Haunebu is to bring the Project Andromeda and Vril VII.

The End of the Reich, the late Saucers?

In August 1941, a RFZ 2 is supposed to have flown to Antarctica, to the German territory of New Swabia. Precisely where the Ahnenerbe and the Thule Society have discovered a passage to a continent under ground, Agartha. Since before the war, the Germans would have built the submarine and underground bases can accommodate their gear.

Some authors have also examined the purported Hitler companions in Antarctica (the Hyperboreans) as well as the links between the Nazi mysticism and energy Vril, hidden civilizations Shambhala and Agartha, and supposed underground bases.

Other writings highlight different scenarios exfiltration of Nazis towards civilizations of the Hollow Earth. In his book Arktos: The Polar Myth in Science, Symbolism, and Nazi Survival ("Arktos: The Polar Myth in Science, Symbolism and Nazi Survival"), Joscelyn Godwin, advance theories about a Nazi survival in Antarctica.

Arktos is known for its scholarly approach and reviewing many sources currently found other than in English translation.
It was also claimed that the PRIL VII "Odin", with a diameter of 120 m, was finally built in great secrecy and it took off from Brandenburg in April 1945 with on board some of the scientists of the Company Vril and Nazi officials after destroying the ground facilities.
After the war, German troops remained in New Swabia would have rallied the Americans, delivering their PRIL type gear, which would have been tested in flight in the California desert.
This is all the more credible that according to some accounts, the Glocke that generated the Vril was also shipped to Antarctica. In 1965, a UFO was observed over several US states. After the supposed crash, residents of Kecksburg reportedly went into the nearby forest in search of remains of what they thought was a meteorite.

On their return, they described a bell-shaped device, with markings on its bottom circle. The army, which was contacted by the sector firefighters,
investing the forest and said he had found absolutely nothing.

A Kecksburg in 1965

Another possibility that the free and spontaneous cooperation Nazi scientists refugees in Antarctica with the Americans is sometimes mentioned. At the end of the war, in 1945, the Americans launched Operation Paperclip to recover, sometimes forcibly, Nazi scientists and their work on the propulsion jet engines, rockets ballistic, etc ...
Richard Miethe, the father of V7, fled Germany in 1945 to travel to the USA. The Russians seized 3 engineers and copies of plans V7 (shared with the Americans). At the Nuremberg trial, the debate on the occult side of Nazism was closed without being opened and nothing filtra on the subject. R. Miethe confirmed in 1953 in the German newspaper "Die Welt" have participated in the development of the famous V7 discs "If flying discs moving in the sky, I pretend to say that they were built in Germany, developed under me, and probably reproduced

in series by the Russians or the Americans. "In terms of scientific refugees in Antarctica, they have can not be worked voluntarily with the US, which then triggered Operation Highjump. VS' was an American operation organized by Admiral Richard E. Byrd in Antarctica. It was launched August 26, 1946 and lasted until 1947. This impressive deployment of forces called in 5,000 men, 13 ships and 26 aircraft.

Three ice breezes Navy deployed at HighJump the operation

The stated objective of this operation was the exploration of Antarctica. However, other projects were carried out in parallel to this, including:

- Do some tests and experiments on the material and on the psychology of soldiers in icy conditions

- Create some basics to establish the sovereignty of

the United States

- Explore the area and make maps All these goals were important, but the most important left to find and destroy any Nazi bases.

The operation itself experienced some official events:

- On 30 December 1946, the aircraft "George I" collided into a mountain during a patrol flight, while photographing the area.
They found survivors two weeks later, but three of the nine occupants of the aircraft were already dead (Wendell K. Hendersin Fredrick W. Williams and Ensign Maxwell A. Lopez).

- The submarine USS Sennett has collided with a large block of ice during a crawl, and had to return to New Zealand. The operation was a fiasco on the line. 1500 people died in the Allies and loss of huge equipment.
In addition much of the photographs was worthless, because in this region compasses were unusable, making it impossible locating shots.

Thus even today the region remains the most unknown of the world, only the coast is clearly delineated on maps, inland remaining unexplored. Most troubling is still probably the Byrd story to his return shipping. In 1947, before starting a new mission in Antarctica, Byrd will say to the press:

"I want to see this country behind the pole. This place is the center of the Great Unknown."
According to official sources, the expedition made a seven-hour flight on February 19, 1947.
But, strangely, he who liked to talk about his exploits to

the press, did not tell anyone about his trip this time. In addition, his behavior changed particularly from that date and no information was available.

It is established fact today, the Pentagon began to military secrecy and confiscated his logbook. For this, it was forbidden to speak in public. After returning from the High Jump operation in the US, another expedition of the US Navy in the Antarctic, with the code name "Operation Windmill", was put on feet.
The objectives put forward by the Chief of Naval Operations for this project was to complement those of Highjump in staff training, testing equipment and reassert American interests in Antarctica.
The expedition was also to investigate the conditions of electromagnetic wave propagation and also to collect geographic, hydrographic, oceanographic, geological and meteorological for information in these areas.
However, although not expressly mentioned in the list of objectives, it was necessary foothold in Antarctica, in order to organize photographic reconnaissance flights. It was mentioned that the 70,000 pictures taken during the High Jump operation had been impossible to exploit, to form cards, because there was no ground for precise registration points (this shows us that the alleged "Mission mapping" was actually hiding something else).

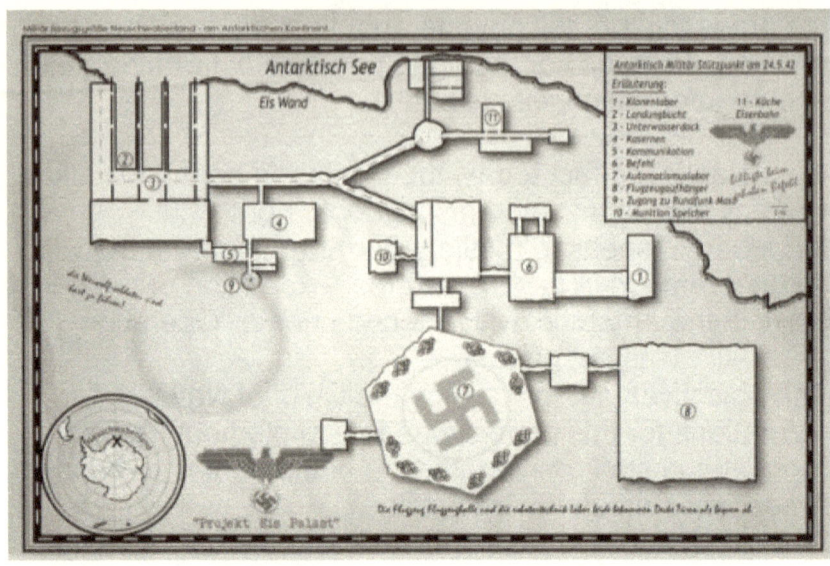

Artist view reconstructing the alleged Nazi secret facilities at the South Pole

If the operation was secretly Windmill end to Nazi activities in Antarctica, it was not a success, since January 8, 1956, several Chilean scientists returning from an expedition to the continent observed for several hours flying objects cigar-shaped and disk in the sky in the area of sea Weeddell.

The same year, 1956, a new military operation will be launched by the Americans: Deep Freeze operation. The result will be even more devastating for the US army will simply then geographic missions and simple recognition to the South Pole.

In a secret concerning two other Deep Freeze operations were launched in the ranks which accounted strategic bombers capable of carrying nuclear warheads.

September 22, 1979, the Vela satellite detected a double feature bright flash of a nuclear explosion. The probes are damaged, it was impossible to accurately determine the location of the explosion.
Yet we know with certainty that the satellite was flying over the south of the African continent, that is, around the New Swabia. If it was an underground nuclear test, it was never claimed by any nation have the bomb. In this context, most likely remains that
Americans, having obtained what they wanted, one way or another, finishing to destroy the evidence of the Nazis passage in Antarctica. It is also curious to note that it is from this date that emerged from the September 22, 1979, the Vela satellite detected a double feature bright flash of a nuclear explosion. The probes are damaged, it was impossible to accurately determine the location of the explosion.

Yet we know with certainty that the satellite was flying over the south of the African continent, that is, around the New Swabia. If it was an underground nuclear test, it was never claimed by any nation have the bomb. In this context, most likely remains that Americans, having obtained what they wanted, one way or another, finishing to destroy the evidence of the Nazis passage in Antarctica. It is also curious to note that it is from this date that emerged from the most documents that allow us to illustrate this.

12 / THE TR3B

The material covered by the aircraft TR-3B is a reactive polymer to the radar, which can change the reflectivity, radar absorption and color. This polymer, when used with some internal electronic systems TR-3B, the vehicle can give an appearance such that it looks like a small boat or on a wheel cylinder on the radar screen.

The TR3B Astra is both a strategic reconnaissance vehicle and a combat vehicle.
A circular ring, filled with plasma acceleration, called magnetic field circuit breaker (MFD) surrounds the rotating compartment of the crew and surpasses all presently known technology.
Sandia and Livermore laboratories developed this secret

technology and government [the US] will make every effort to protect it. The plasma generated by increasing the mercury 250 000 atmospheres and at a temperature of -123 degrees Celsius, while it is accelerated to 50,000 rev / min becomes a superconducting plasma, this process, with the ultimate result canceling the gravity (levitation).

Specifically, by generating the rotating magnetic field, the effect of the gravitational pull on the ship is neutralized to 89%. In other words, the weight of the circular accelerator and all the weight of it - Crew hood, avionics (navigation devices and devices and operating), fuel systems for environmental the crew and nuclear reactor, etc. - are reduced by 89%.

The TR-3B aircraft operate at high altitude undetectable manner (with STEALTH technology), providing a platform for recognition of time indefinite flight.
Once it reaches (high-speed) at that altitude, it only needs very little propulsion to maintain altitude.

Also, I remember that there were rumors of secret Groom Lake base for a new element capable of playing the role of plasma catalyst.
With reduced weight by 89%, the vehicle can run at a speed of Mach 9 [9 times the speed of sound: approximately 10 000 km /h], both vertically and horizontally.
Sources claim that performance is limited only by the biological limitations of the drivers, which are many, because the gravitational force is reduced by 89%.
Propelling the TR-3B is provided by three multimodal propellers mounted in each corner at the bottom of the triangular platform.

The TR-3B moves at speeds up to Mach 9 (Mach is a unit of measure used in aerodynamics to express the speed of a moving body in a fluid: projectile, airplane, rocket, etc.
The Mach 1 speed is equal to the speed of sound in said fluid;
Under normal conditions, Mach 1 is equal to 1224 km / h (or 340 m / s) up to 120 000 m (120 km), then God knows how fast it can move.
The three multimode rocket engines mounted at each corner of the platform use hydrogen or methane and oxygen for propulsion. In a fuzing system using hydrogen and liquid oxygen, 85% of the propulsion supplies oxygen.
Nuclear rocket engine uses a hydrogen booster and liquid oxygen for more power.
The reactor heating liquid hydrogen and liquid oxygen injected through supersonic nozzles, so that hydrogen burns simultaneously with liquid oxygen. Multimodal propulsion system can be operated in the atmosphere with the propulsion provided by the nuclear reactor in the upper atmosphere with hydrogen drive and the combined

circumferential orbit orbit (hydrogen / oxygen).
We must not forget that the three rocket engines propel only 11% of the weight of the airplane Top Secret TR-3B. The engines are built by Rockwell. Therefore, when the triangular UFOs are saved, this does not necessarily mean they are of extraterrestrial origin, and can also be ultra-sensitive theft devices, such as the TR-3B.

The National Security Agency (NSA), the National Reconnaissance Office (NRO), the CIA and the United States Air Force (USAF) intentionally created some ambiguity concerning the names of these aircraft: they created the TR3, modified TR-3A, TR3-B plus 2, 3 or 4 suffixes, meaning in fact completely different aircraft

Some of them are designed to have a human crew, others are designed to operate without a pilot. " Independent journalists who courageously reveal all this information secret hope that their appearance on more and more media will greatly help "awaken" the people face these hidden realities and bring them and to use them for the good of all humanity. Moreover, the disclosure in the press of this information will prevent and completely dismantle the various manipulations of a possible alien invasion or similar rumors trying to escape the "black operations" satanic elite self – called "enlightened" ".

13 / OTHER ANTI GRAVITY SPACECRAFT

Ten projects have emerged in the USA. It is :

The stealth bomber Northrop Grumman B-2 Spirit

The Aurora

The Lockheed Martin X-33A

The Lockheed X22A

The Nautilus Boeing and Airbus Industries

The TR3-A

The TR3-B Astra triangle

The DARPA Falcon

Disc Great Pumpkin Northrop / USAF

XH-75D Teledyne Ryan Aeronautical

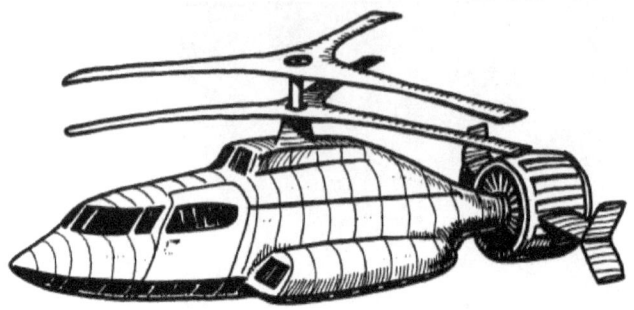

It there's different forms of generation antigravity fields.

The most primitive antigravity technology is electro-gravitic.
This involves the use of millions of volts tensions to disrupt the ambient gravitational field. That translates into a 89% reduction of the resistance of aircraft in vehicles such as the Stealth Bomber B-2 and the TR3-B triangular craft Astra.

The next level of sophistication is magneto-gravitic.
This involves the generation of high energy toroidal fields spun at phenomenal speeds, which also disrupts the ambient gravitational field and even the force generated by the force of the gravitational pull of the Earth.
The third level of sophistication, the one used in most modern American craft antigravity is the direct generation and exploitation of the powerful gravitational force.
Such a field of strong force extends slightly beyond the atomic nucleus Element 115. By amplifying this exposed gravitational force, and using the high-energy reactor antimatter, then the leader, it is possible to raise a craft from Earth, then change direction by vectorizing the force field antigravity shaped thus generated.

14 / THE USO

OSNI means underwater object unidentified. Unlike flying saucers, there is very little physical indications of the USO, such as photographs or motion pictures, even if they are distant, vague and lacking in detail, as is usually the case with unidentified flying objects .

The military reports on any submarine detection can not be very precise, the nature of the aquatic environment by preventing. However, as for air events, there is no evidence of their true existence.

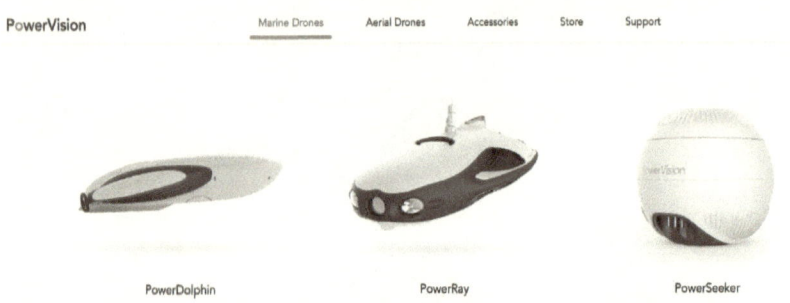

And because most of the USO can be seen at sea are actually underwater drones.

PowerVision PowerSeeker Fishfinder

Some are used for monitoring and recognition as those of

PowerVision (see photo above) and others are purely military and may be armed, they are then more missiles such as the XFC UAS.

Launch of XFC UAS lapse photography. Deployed from the submerged submarine USS Providence, the plane without pilot developed XFC by the NRL is launched vertically from a launcher 'Sea Robin' (bottom right). The UAS folding wing deploys independently aerodynamic profile X-wing and, after reaching a marginal altitude, adopt the configuration horizontal flight. (NAVSEA-AUTEC)

The XFC UAS is a fully self-folding wings, powered by a fuel cell with superior battery life to six hours.
The unhybridized panel supports the propulsion system and payload for endurance theft for ISR missions relatively low cost and low altitude.

The XAS UAS system uses an electrically assisted takeoff system that lifts the plane vertically out of its container and therefore allows a very small launch, for example, from a van or a small surface vessel.

15/ 70 YEARS OF PSYCHOLOGICAL WARFARE

The term "UFO" was coined in 1953 by the US Air Force (USAF) to serve as a catchall for unidentified flying objects reports.
Curiously, the origin of the modern UFO history seems to correspond to the creation of the CIA. The government should have conducted investigations and studies, many of which were inconclusive. However, CIA documents emphasized the utility and use the phenomenon to test the crowds.
The CIA suggested that the agenda behind the emergence of UFO and alien stories is a long-term PSYOP and multifaceted:
voltage
1.Stratégie - finance science fiction movies and video games to prepare for a war.
2.Develop magical thinking and hysteria or what Winter

Watch called "cartoon world".
3. Programmation neurolinguistic to consolidate the "conspiracy theorists" in a register for the purpose of discrediting.
4. Distraction or wrong direction - In 1973, a survey revealed that 95% of the public had heard of UFOs.
5. Jeter the groundwork for a false flag project (under false flag) called Project Blue Beam.

The most commonly identified sources of UFO reports are actually:

• Astronomical objects (stars, planets, meteors, secret spacecraft, artificial satellites ...)
• Aircraft, military vehicles, missile launches
• Balloons (weather balloons, large research balloons)
• Other atmospheric and meteorological objects and phenomena (chemtrails, kites, flares)
• Light phenomena (mirages, Fata Morgana balls flashes, projectors and other floor lighting)
• Hoaxes

It there's little the CIA of herself admitted that she was responsible for the majority of reports of UFOs during the Cold War with Russia.
"It was us," the agency published on social media, accompanied by a document declassified in 1990.
The 272-page report, strongly written entitled "The CIA and the U2 program, 1954-1974," describes various aspects of the U2 CIA program that was used to high altitude spy missions into space Russian air.
The report of the 1998 CIA continues by revealing how the deployment of the U-2 aircraft with its revolutionary capabilities virtually fueled hysteria of UFOs during the Cold War. What the report does not say explicitly is that the CIA to make UFO hysteria.

The secret U-2 surveillance program was essential for the Agency, and the report discussing various "cover articles" used to conceal the program and its purpose. The myth extraterrestrial spacecraft visiting Earth was a perfect diversion to allow the public imagination worsen to the point that prevented the Russians from knowing the real US capabilities. Psychological operations are one of the main functions of the CIA, after all, the agency manipulates and influences the American public for over 70 years! Secret Notes forged by the Office of Special Investigations of the US air force would have been disclosed to ufologists that aired to the public the conspiracy theory to disguise the government's actions. The method, called "perception management ", designed to distract people from the reality of the true paradigm.

15.1 / RICK DOTY

A fascinating film has treated the subject of perception management.
This is " Mirage Men " by John Lundberg from the book of the same name from the English author Mark Pilkington. The documentary focuses on the United States, particularly a handful of characters and well known in UFO circles events.

Richard Doty, former officer of espionage against the Office of Special Investigations of the US Air Force (AFOSI), is a key player in the book and in the film. Another is ufologist and author William Moore, whose books include " The Roswell Incident " and " The Philadelphia Experiment ".

Rick Doty, as he calls himself lately, and Moore seem plausible enough to start in front of a camera, although they share a dark past.

Both acknowledge having provided false information to one Paul Bennewitz, a veteran of World War II and electronics expert installed alongside a USAF base in New Mexico.

According to Doty, the NSA has targeted P.Bennewitz in the early 1980s for use in their black propaganda wars, sending false messages extraterrestrials in his house and planting false evidence spaceship on a nearby mountain. Convinced that he had discovered an alien invasion plot, Bennewitz began writing letters of warning to President Reagan, has developed a paranoid psychosis and was briefly institutionalized. Mission accomplished.

" Mirage Men " began a fascinating excursion into a twilight zone of wild conspiracies, haunting believers and double Psychological agents, even though the claims and requests never lead a true investigative journalism.

Indeed, Pilkington does interrogate Rick Doty explains that he has manipulated Bennewitz because he saw it was not to see; A secret US technology. You had to believe him at all costs that what he saw was well indeed when he had absolutely nothing.

But more than just a desire to hide anti-gravity devices must also make the public believe that extraterrestrial life is possible, although officially as soon as there's an observation we did not necessarily explained.

We must leave some doubt this unhealthy because never,

ever the theory of evolution should be questioned.
We can not explain to people that by the end of the 50 atmospheric tests were carried out and revealed that we live in a closed system.

I will not say more in this book not to shock some people and I
invite anyone asking more questions to go to my website: https://cyprustar.wordpress.com/.

15.2 / STEVEN GREER

Steven Greer, born June 28, 1955 in Charlotte in Mecklenburg County in North Carolina, is a physician, ufologist and essayist. It has embarked on a vast communications operation at international level, producing a film called "Sirius", officially presented in Los Angeles on April 22 2013, quickly released on DVD in June 2013. After fighting for the thesis that the UFO phenomenon corresponds to alien visits, Greer explains that US government would have tapped into the UFO issue, essential information that would have allowed him to dominate a technique to extract vacuum energy ("Zero Point Energy").

An energy that could save humanity altogether. But the forces of darkness would retain the secret to better secure the population under their yoke.
Only if Greer had any evidence of its alternative energy sources, why not show some?
I bet if he had an interesting proof, it would be able to raise $ 1 million in no time.
The fact is that the Orion Project Steven Greer is a big joke.
In the film "Sirius" you see people who follow one of his lectures, enter one by one in a room.
Wizards inspect their body with a metal detector, similar to what is used in the vicinity of gates at airports. And Greer explains that his approach is so daring that he must protect his life. It was, he said, threatened with death. It shows Greer animating vigils and say, "If you put yourself in meditation and deep down in your heart you wish ardently that the UFO phenomenon is manifested, then your wish will be granted."

Greer says that the US government has been able to appropriate the technology used by the aliens. It master a new form of energy, zero-point energy. If these secrets were revealed to the world, our humanity could experience a golden age. But the powers that power, would lose in the process. So there is a great conspiracy that the secret is well kept.
This is the point of the film it will be relatively easy to disassemble, as evoked fixtures are grotesque. But the American public is gullible, and the general public around the world, probably too.
People want to believe it.
In the sequel Greer casts a wide net, will allow all possible speculations. Anything goes: doubts about the official version of September 11, manipulations financial, the FED, occult groups, etc ... His talent is to mix the true with the false, which I grant you is not always easy to disentangle.
Amidst all this, the team has a be 15 cm long, recovered, dried in the Atacama desert, by a person whose Greer can not reveal the identity.

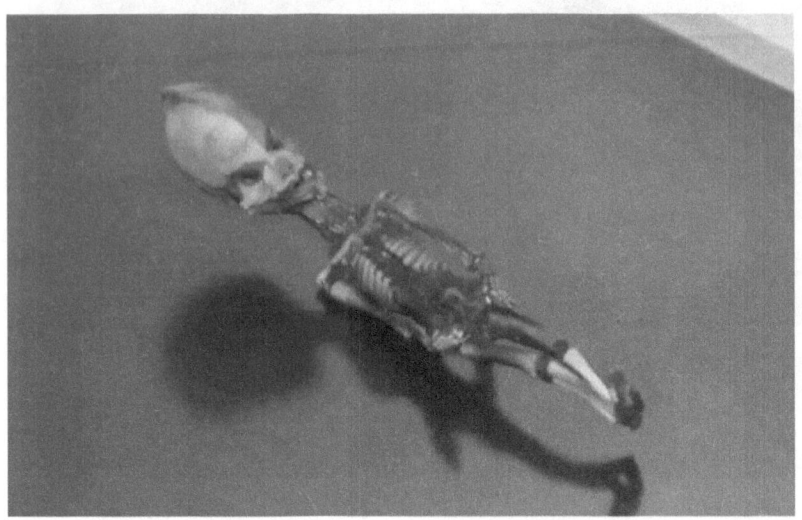

The star of the documentary is Ata. Since the posting of teasers documentary photo of the creature made the Anglo-Saxon newspapers and trailer promises' historical results. "

However documentary disappoint many fans of the paranormal as scientific analyzes provide explanations contrary to those expected by ufologists: DNA Ata is human.
The analyzes presented in the documentary is the work of Gary Nolan, director of the stem cell department at the medical school at Stanford University, who conducted the analysis in the fall 2012. He believes that the death was about a century.

End of September 2012, reviews are carried out in Barcelona in Spain.
DNA analysis confirms that this is a human, the sequences obtained being closer to man than the chimpanzee.

So one wonders what Steven Greer wanted to do with his Sirius movie!

15.3 / OTHER ALIENS HOAXES

1953: Spaceman

Three young men in Georgia would have told police they had hit a van that looked like a small creature from outer space.
A local veterinarian confirmed this to be round ears was not a known animal humanity but anatomists from Emory University who studied the body disagreed: The alien was actually a capuchin monkey with shaved tail removed.
The three men confessed to the hoax organized to appear in the local newspaper.

1995: Autopsy " Roswell "

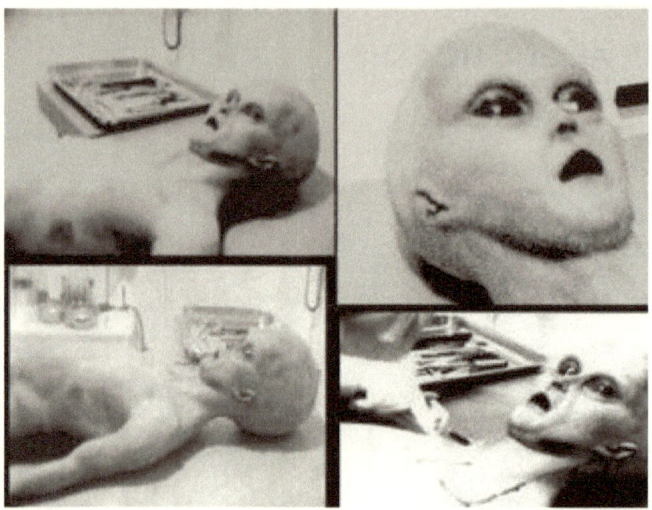

Supposedly a film physicians practicing a particular autopsy on an alien killed in the Roswell crash in 1947, the film turned out to be exactly what it looks like: a hoax using a rubber dummy, animal parts and raspberry jam.

1996: Alien Dr. Reed

A Dr. Jonathan Reed stated that the body of this alien being had been stolen by government agents who continued to stalk him and threaten him (although they forgot to confiscate his pictures of the UFO and extraterrestrial frozen).
Finally they learned that Jonathan had created a false identity and his name was Jonathan Bradley Rutter. He refused to show his " evidence ".

1999: The Starchild Skull

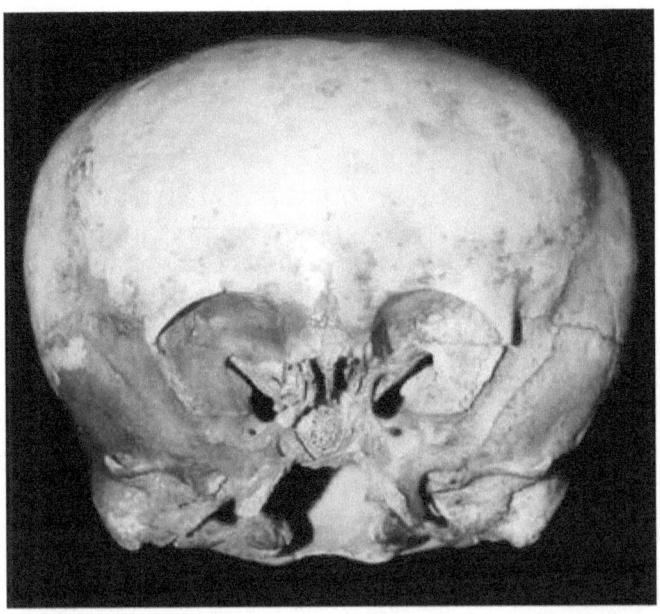

In 1999, the American novelist Lloyd Pye bought what turned out to be the skull of a hydrocephalic child. but it will claim that this is an alien hybrid.

2005: Yugoslav Autopsy

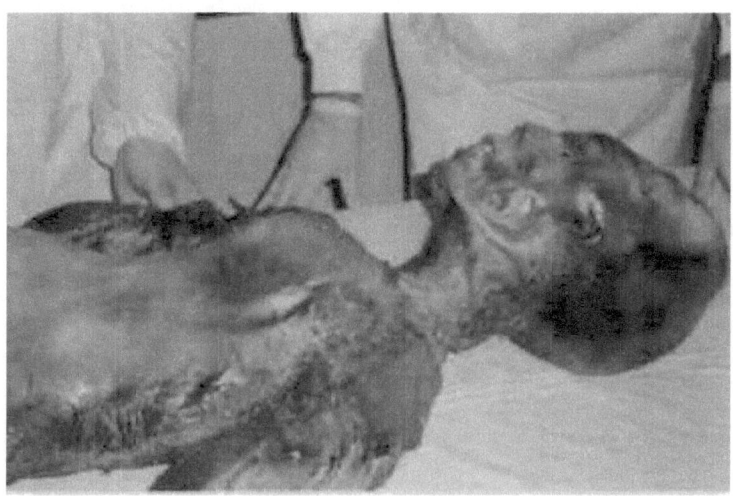

This film was shot in the former Yugoslavia 1966. On the photos sent to UFO Casebook by Ivan Kremer, we see doctors Andrej Zobol and the two pathologists Nikola Jullic and Zoran Frederic examine the charred body of an alien from a crashed UFO in the village of Otocek. But it is the doctors who are quoted are fictitious and no one could find them in the phone book, not even those who had access to military directories, which confirmed the hoax.

2006: Alien in a jar

During the renovation of a cottage Gunthorp, workers found a jar containing what was actually a realistic alien model made of clay.
Who had put the alien model in the attic of Barney Broom and why it remains a mystery.

2011: Alien Siberia

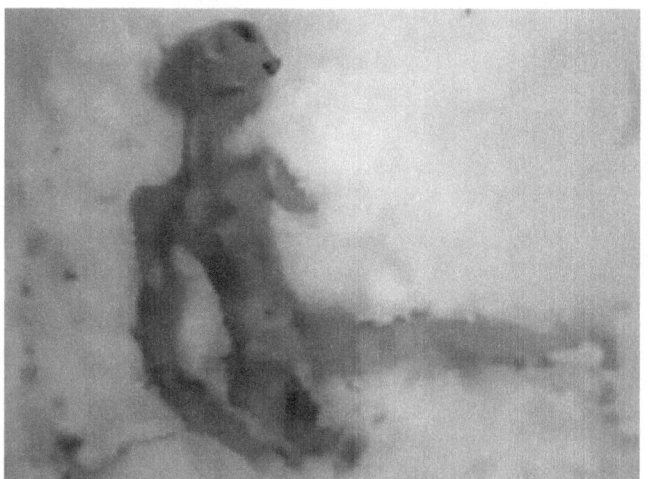

Worldwide media have broadcast the video filmed with a mobile phone and posted on YouTube by anonymous teenagers, showing an alien entity pitifully lying in the snow. The Kremlin wate an investigation, and a few hours later we learned that the alien in question was actually made from old bread and chicken skin.

16 / THE SHADOW GOVERNMENT

Here is a summary report on the elements of this clandestine organizational network we call the shadow government, which serves as a kind of "parallel government" in the official government elected and appointed US.
It includes elements known Richard Bylan (character which I grant credit only on certain points and it is a part) with sufficient certainty to be identified positively, and their known or described functions reliably are described.
It is quite possible that there are other elements (particularly in the areas of "black budget" and "special operations") that escaped his study and are not named here.
However, unlike the official government, non-executive branches of shadow government simply intended to distribute various functions, but not to set up a system of counterweights, as was supposed to happen constitutionally between the executive, legislative and judicial.
In the shadow government five branches can be identified:

- executive power,
- the part of the information,
- the department of war,
- the arms industry branch, and
- the financial department.

An analysis of the overall objectives of these five branches suggests that the overall purpose of the shadow government is to exercise a secret inspection:

1. Collect comprehensive institutional and personal information

2. Establishing a national and independent international policy of the incumbent government

3. In developing weapons and high-tech equipment and, with these, establishing small military units specialized elite, highly mobile and highly mobile, to apply these secret political, if necessary, without having to rely on the armed forces (and "unreliable") services (including subjection to phantom Government is reasonably suspect)

4. By developing an armed capability to repel any threat using BMDO based surveillance and SDI weapons network

5. By denying information compromising the shadow government of all the people outside the decision-making levels "need to know"

6. By controlling the money supply, the availability of credit and the value of money, through policy decisions made outside the official government.

These control mechanisms are used to preserve or advance the agenda of an international group of power brokers and influence.
According to Senator Barry Goldwater, the agenda states that "national boundaries should be erased and world rule established." The instrument unifying the most visible of these is power brokers,

• The Council on Foreign Relations (CFR) (promoting the transition from the Earth a group of nation-states to a world government), [Chairman: Peter G. Peterson; Headquarters: 58 E. 68th Street, New York, NY 10021].

• However, do not underestimate the influence of the Trilateral Commission (TC) (coordinating economic initiatives Group of Seven with other "developed countries" vis-a-vis the "underdeveloped world")

• One should not ignore the power of the secret Bilderberg Group (BG) (which focuses on the considerations Military and Strategic powerful Western Europe and brokers in North America), [the President seat by rotation: the former president: Prince Bernhard of Holland.

David Rockefeller is the Chairman Emeritus of the CFR and the TC,and certainly influenced by through representatives proxy (such as Lloyd Bentsen), the Bilderberg group. The following is an identification and brief description of the constituent bodies of each descinq branches of the shadow government.

EXECUTIVE MANAGEMENT

This branch includes the decision-making structures and effective control behind the veil of apparent democratic government structures:

a. Council on Foreign Relations (CFR)

b. trilateral Commission

c. The Bilderberg group (Prince Hans-Adam of Liechtenstein, Prince Bernharddes Netherlands, Bill Clinton, Lloyd Bentsen, etc.)

d. National Security Council (NCS) (underground group of the CIA), its secret 5412 Committee and its PI-40 Subcommittee (eg Majestic 12);

e. The Fund for Special Operations Joint Chiefs of Staff (JCS) (the lead which implements the orders of the 5412 Committee of the CSN, with the command of US Special Forces);

f. National Program Office (NPO) (which manages the project Continuity of Government (COG)), a secret project underway to maintain command centers, control, communications and intelligence during a national emergency extreme leveraging secure underground cities and clandestine substitution with staff to national leaders off-ground.

g. The fund black projects of the Federal Agency for Emergency Management (FEMA) (which manages federal custody camps [often located on military bases or on land the federal Bureau of Land Management] secures shelters underground for the elite during cataclysms, etc.).

DIRECTION OF INTELLIGENCE

Assume national and international monitoring functions and secret police / performers:

a. The National Security Agency (NSA) (monitors and controls all phone calls, telegraph, computer modems, radios, televisions, cellular communications, microwave and satellite and electromagnetic fields "of interest" around the world, and band control and activities related to UFO secrecy and surveillance operations of aliens), Fort Meade, MD;

b. The National Reconnaissance Office (NRO) (control and information collection global spy satellites, coordinates arms fire to beam energy from satellites orbiting satellites focused on human or airborne targets and secret devices that passing for aliens) the Pentagon basement and Dulles Airport area, VA;

c. The National Organization Recognition (NRO) (aka MJ-TF) (the branch operations of intelligence / military PI-40 Subcommittee) monitors and "interacts" with people close to the ET meeting, including occasional kidnappings control of the attackers mind, physical and sexual, disguised as "alien abduction" in war purposes psychological and purposes of disinformation), unknown headquarters, probably compartmented and dispersed among various elite units of the Special Operations Delta Force, as the US Air Force Blue Light at Hurlburt Field, Mary Esther, FL and base Beale air, Marysville, California;

d. Central Intelligence Agency (CIA), (command, control often and sometimes coordinates the collection of secret information abroad gathered by spies (HUMINT), electronic surveillance (SIGINT) and other means; running covert operations of against-insurrection against paramilitary unconstitutional international law, as well as operations against espionage against foreign agents, national surveillance and manipulation of the American political process, "in the national interest" in direct violation of the Charter of Congress, and operates "false front" companies owners a significant portion of international transshipment of illegal drugs, using the cover of national security and immunity, and cooperates with NSA '

e. Department of Energy before (DOE- INTEL) (which performs security checks internal and external security against-measures, oftennthrough its contract civilian authority, the Wackenhut Corporation);

f. The central security service of the NSA and the Special Security Bureau CIA (who spy respectively spies and conduct special operations which can not be entrusted to line intelligence officers), Fort Meade, MD and Langley,

g. US Army Intelligence and Security Command (INSCOM) (whose missions include psychological and psychotronic warfare (PSYOPS), the parapsychological intelligence (PSYINT) and electromagnetic intelligence (ELMINT), Ft Meade, MD.

h. Office of Special Investigations of the Army US Air (AFOSI) (which gathers information about aerospace operations and has a compartmented unit involved in investigating UFO sightings, contact reports " aliens ", and monitoring by the IAC [Identified Alien Craft] and the coordination with the prohibition operations NRO), Bolling air Force base, MD;

i. Defense Intelligence Agency (DIA) (coordinating intelligence data collected from various intelligence services of the armed forces (army, navy, navy, air force, coast guards and special forces) and provides measures against the threat (including security). in ultra-classified installations by the deployment of the "thought police" American, which monitors, remote and other parapsychological measures, penetrations and scans performed by viewers, Pentagon, VA, Fort Meade, MD and the entire astral plane;

j. The Defense Industry Security Command (DISCO) (which conducts intelligence operations within and on behalf of civilian defense contractors engaged in research, development and production classified);

k. Defense Investigative Service (DIS) (which investigates people and situations considered a potential threat to any operation of the Department of Defense);

l. Naval Investigation Service (NIS) (which investigates threats to naval operations);

m. Informations the fight against drugs agency (DEA) (which monitor and prohibit drug smuggling operations, unless otherwise provided for by the derogation "National Security");

n. Defense Electronic Security Command (which coordinates intelligence surveillance and threat prevention measures to the integrity of military electronic equipment and electronic battlefield operations), Fort Worth, TX.

o. Project Deep Water (lingering effects of staff, sources and methods compromised resulting from the secret importation of General Reinhard Gehlen, head of the Nazi intelligence to Hitler, to redefine the American intelligence apparatus);

p. Project Paperclip (the ongoing results of the secret importation of Nazi weapons armament and aerospace scientific bases in research and secret military development in the US);

DEPARTMENT OF WAR

Development of high-tech weapons and deployment of special units of Special / Special Operations Forces:

a. The Directorate of Science and Technology CIA (which brings together promising information on scientific and technological developments of superior advantage on national security or a threat against national security)

b. Office of the Strategic Defense Initiative (SDIO) / Ballistic Missile Defense Organization [sic] (BMDO), which coordinates the research, development and deployment of the satellite weapons

c. Department of Energy (DOE) (Department of Energy), which in addition to its feature article on the research of coal and less polluting and more solar energy essence, is mainly involved in research and development on: the more specialized nuclear, compact, autonomous and operated fusion, particle and wave weapons, including pulses electromagnetic, applied research on gravitational weapons / antigravity, laser, particle beam and widgets, technology "Masking" invisibility by high-energy, etc.)

d. Lawrence Livermore National Laboratories (LLNL) / Sandia National Laboratories-West (SNL-W) (participating in the "refinement" of nuclear warheads, the development of new trans-uranic elements for applications in the field of weapons and the energy, development of weapons release material (Teller bomb: 10,000 times the force of a hydrogen bomb), applications of laser technology / maser, Livermore, California

e. The national laboratories engineering Idaho (INEL) (which are home to many underground facilities in a complex of facilities in the vast desert greater than Rhode Island) benefit from the security provided by their own secret naval base part for nuclear research, electromagnetic high-energy, etc. includes Argonne National Laboratory, West), Arco, ID;

f. Laboratories Sandia National Laboratories (SNL) / Phillips Air Force (sequestered in the military reserve Kirtland Air Force / Sandia) and translate research into theoretical and experimental research on nuclear and nuclear weapons and Star Wars (satellite) conducted at Los Alamos and Lawrence Livermore. practical and functional weapons), Albuquerque, New Mexico

g. Tonopah Test Area (DOE testing center for testing of SNL operably Star Wars weapons in realistic target situations, and is adjacent to a stealthy and hidden aerospace base and US-UFO bases at basic Groom Lake [USAF / DOE / CIA] [Area 51] and the base of the Papoose Lake [S-4]) test site in Nevada / range of Nellis air Force base, Tonopah, NV

h. Haystack (Buttes) Laboratory of the US Air Force, Edwards AFB, CA (extreme security center to a depth of 30 levels have been involved in reverse engineering

i. Los Alamos National Laboratories (LANL) (first research laboratory on nuclear technologies, subatomic particles, high magnetic fields, exométallurgiques technologies exobiological and other exotic technologies), Los Alamos County, NM

j. Area 51 (Groom Lake base [USAF / DOE / CIA]) and S [site] -4 (basic Papoose Lake) Ultra-secure deployment bases "nonexistent" in which extremely classified aerospace vehicles are tested and flown operationally, including the Aurora [Mach-8] a spy plane, the stealth fighter Black Manta [TR-3A] succeeding F- 117A, aerospace reconnaissance vehicle unmanned hyperspeed hypervironique, and several variants of anti-gravitational vessels

k. US Special Forces Command, Hurlburt Field, Mary Esther, Florida, with his headquarters in the western United States Special Forces Command, Beale AFB, Marysville, CA, coordinating: 1) US Army Delta Force (Green berets), US Navy SEALS (black berets), Coronado, CA) strike force of the US Air Force Blue Light (red berets)

l. Research Projects Agency for Defense Advanced (DARPA) (which coordinates the implementation of the latest scientific findings in the development of new generations of weapons); (Now called ARPA)

m. Jason Group (scientific elite in weapons applications, developing weapons at the forefront of science for DARPA / ARPA and operating under the cover of the Miter Corporation); not. Aquarius Group, an elite group of scientists aware of highly classified scientific and technological discoveries

o. Defense Science Board (used intermediary between weapons needs and the physical sciences of the Department of Defense)

p. Defense Nuclear Agency (DNA) (currently focusing on development and deployment of weapons of field-strength-laser type, x-ray beams of high energy and high-energy particles)

q. US Space Command, (headquarters of space warfare to lead "the next war will be fought and won in space"), Falcon AFB, CO

r. North American Aerospace Defense Command (NORAD) (the operator of the space monitoring center and war command with a nuclear survivability in the heart of Cheyenne Mountain), Colorado Springs, Colorado

s. Office space systems of the Air Force (which coordinates the development of future technology for operating and fighting in space)

t. Project Cold Empire (classified research on SDI weapons)

u. Project Snowbird (pseudo-UFO's used as disinformation)

v. MILSTAR project (development and deployment command satellites, control, communications and intelligence of World War III [space war])
w. Projet tacit rainbow sky (stealth drones / pseudo-UFO)

x. The Timberwind project (spacecraft nuclear-powered)

y. Project Cobra Mist (SDI weapons research energy beam)

z. Project Cold Witness (classified for SDI weapons), etc.

BRANCH ARMS INDUSTRY

a. AT & T (Sandia Labs, Bell Labs, etc. - Research on Star Wars and weapons facilitation interception telephone / satellite NSA); (The Batia Weapons Lab was taken over by the Battelle Memorial Institute, a proprietary account with declared intelligence links)

b. Stanford Research Institute, Inc. (SRI) (subcontractor intelligence involved in research psychotronic, parapsychological and PSY-WAR)

c. RAND Corporation (CIA group involved in Intelligence projects, weapons development, and underground bases)

d. Edgerton, Germhausen & Greer Corporation (EG & G), (NSA contractor / DOE involved in the development of weapons Star Wars fusion applications and security for Area 51 (basic aerospace vehicles anti gravity technology) and nuclear facilities

e. Wackenhut Corporation (NSA contractor cutting / CIA / DOE) involved in contract security operations for underground military reservations and underground Top Secret Ultra and Black Budget, such as the S-4 area (US UFO base), NV and Sandia National Labs (Star Wars weapons base), NM) and seem t- it, of "dirty jobs" for the CIA and the defense intelligence services

f. Bechtel Corporation ("ditch digger" of the CIA for secret projects and underground bases not provided);
g. United Nuclear Corporation (military nuclear applications)

h. Walsh Construction Company (on CIA projects)

i. Aerojet (Genstar Corp.) (combat satellites manufactures DSP-1 Star Wars NRO)

j. Reynolds Electronics Engineering (the CIA / DoD)

k. Lear Aircraft Company (Black budget technology);

l. Northrop Corporation (makes a device anti-American seriousness, manufactured near Lancaster, California)

m. Hughes Aircraft (classified projects compartment);
not. Lockheed Martin Corporation (Black Budget aerospace projects)

o. McDonnell-Douglas Corporation (Black Budget aerospace projects)

p. BDM Corporation (subcontractor of the CIA involved in back-engineering and psychotronic projects)

q. General Electric Corporation (electronic warfare and weapons systems)

r. PSI-TECH Corporation (involved in military applications /intelligence research in parapsychology)

s. Science Applications International Corp. (SAIC); "Black projects" contractor who apparently understand the psychological war.

FINANCIAL DEPARTMENT

Extra-constitutional funding:

a. Federal Reserve (cartel of private banks overseen by the super-rich financial elite, such as the Rockefeller, Mellon, the DuPonts, the Rothschilds)

b. CIA self-financing (operating and / or control of much of the international drug trade with heroin, cocaine and marijuana, as well as "front" companies, as a source of money for unofficial covert operations, purchase of exotic munitions and strategic wine pots de-fund)

c. Self-financing of the Ministry of Justice (use of confiscated cash and valuables from "targets of investigation" to finance "special projects")

d. Special forces are self-financing (using the "booty" confiscated during covert military operations to fund other clandestine operations).

The shadow government is a big parallel power structure, perfectly organized and well camouflaged.
The story suggests that he has served his masters and his predilection to act away from the sight and reputation, if not totally illegal way, exactly the will of his masters : do not draw attention to himself and accomplish through covert operations which can not be accomplished in legal or political basis.
What should be the attitude of informed citizen vis-à-vis the shadow government?

Since thriving in the dark, we should shine the light of full disclosure on this. Citizens may require:

• The end of the practice of Congress to allow Black budget;

• The end unpublished secret decrees and national security directives;

• The end of the practice of presidential declarations of national emergency maintained indefinitely (as currently in force);

• The end of the government's domestic spying of its own citizens;

• The extremely severe reduction (approximately 90%) in the number of staffing and scope of the intelligence agencies which proliferate continuously, which are an anachronism since the end of the cold war and end the collusion between the CIA and the DEA allowing a continuous flow of drugs from spreading in the country.

17/ ELECTRIC SENSIVITY TEST

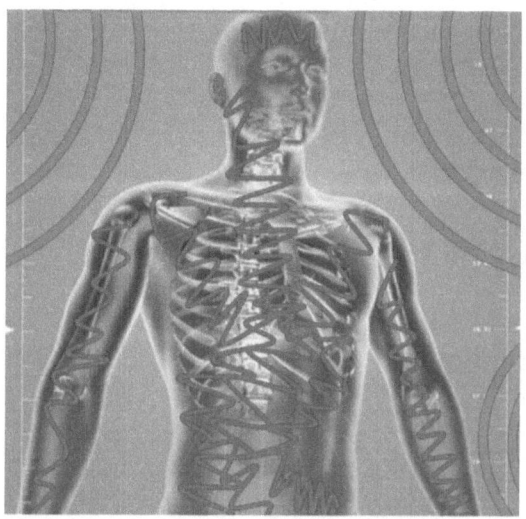

Although these are not extraterrestrials who are at the base of UFO sightings around the world, gear to the source of the sometimes traumatic events for their witnesses can cause injury visible or not. Those invisible electromagnetic waves come they can send. Personally I have been a victim and thirty years after my experience with the TR3B I can still cause electrical abnormalities around me.

All these issues gives a general idea of the level of sensitivity of the subject to electromagnetic phenomena, to assess if it is above average.

1 / Are your eyes are particularly sensitive to light, noise, vibration and the crowd?

2 / Fluorescent tubes do they tire or do they make you sick?

3 / they seem Light bulbs last for a short time at home?

4 / Electrical appliances do they turn on and off they sometimes alone at home?

5 / Quartz watches or other kinds of watches do they hurt when you wear on your wrist?

6 / Do you feel uncomfortable in synthetic clothing and do ye that natural materials?

7 / Do you sometimes have the certainty that someone or something is in the room with you, while you do not see anything special?

(This effect can be reproduced by applying a magnetic field to the brain. [Research L. Ruttan, Persinger and Koren S.].

8 / Are you allergic to city gas?
(For Albert Budden, allergies are linked to electromagnetic hypersensitivity.)

9 / Are you allergic to perfume to aftershave, makeup, gasoline, the smell of paint, aerosols, etc.?

10 / Are you allergic to some food or some drinks?

11 / Are there sometimes periods of the day on which you have no memory of what you did?

12 / The time you he sometimes seems slow or fast flow?

13 / Do you sometimes feel painful electric ripples under the skin?

(This is called fasciculation.)

14 / Do you notice all your experiences or do you write a lot?

(Albert Budden found a taste for writing in "electric sensory", often on religious themes or have a cosmic dimension. Some subjects also reported the ability to "speak in tongues", well-known phenomenon of renewal of communities charismatic, and could be due to a disturbance of electromagnetic origin affecting the language areas (Broca and Wernicke areas).

15 / Do you often have a feeling of "déjà vu"? 16 / Do you see balls lights at home while the others do not see?

(This type of hallucination caused by an electromagnetic field has been reproduced in the laboratory.)

17 / Are you particularly sensitive to lightning storm or the passage of a plane?

18 / Have you ever been close to the point of impact of a thunderbolt, was close to a lightning ball or was electrocuted (s), etc., when you were younger?

If not, what happened to your mother when she was pregnant with you?

(Albert Budden adds to this list the case defibrillation (cardiac resuscitation) and electroconvulsive therapy.)

19 / Were you a premature baby?

(Incubators typically produce an electromagnetic field that could create awareness in children. [C.Smith, and R.Choy

J.Monro].)

20 / Do you have a tendency to diabetes or hypoglycemia?

(According to Albert Budden, pancreas would be particularly sensitive to electromagnetic radiation that could disrupt its operation.)

21 / Do you live near an electric pylon, power line, radio-transmitting antenna, a telecommunications tower, an electrical transformer, etc.?

22 / Do you live near a river or an underground river, or near a geological fault?

(Albert Budden adds to this list the military areas because radar and radio broadcasts. It empirically considers the critical distance is 450 to 750 m between the electromagnetic source and home of the subject. However, in Paris, the thousands of people living near the Eiffel tower on top of which is a powerful television transmitter without being inconvenienced. This matter therefore has its limits and called a check using measuring devices the presence of electric fields and magnetic assumed.)

23 / Do you live experiences "paranormal" (precognition, telepathy, clairvoyance, gift of healing, psychokinesis, disembodiment, etc.)?

24 / Did you have a happy childhood?

(Kenneth Ring spoke with victims of EMI or RR4, the possibility of being an abused child or a victim of sexual abuse.)

25 / Do you take frequent discharges of static electricity by touching a doorknob, a car, etc.?

26 / Do you sometimes feel a metallic taste in the mouth?

(This taste may be due to the presence of an electromagnetic field acting on dental fillings in the mouth.)

27 / Objects Disappearing sometimes you, or objects behave strangely?

(This question refers to poltergeist phenomena of electromagnetic origin: disappearance of objects as a result of an unconscious movement by the subject when it suffered a temporal lobe epilepsy, psychokinesis or levitation.)

28 / Have -you ever felt complete silence and abnormal around you?

(This effect is due to an electromagnetic field acting on the brain of the subject and is sometimes reported facing a UFO appearance. It may be reproduced in the laboratory [see work R.Thompson].

18/ DIGITAL SOURCES

1/ UFO IN THE ART AND HISTORY?

http://www.sprezzatura.it/Arte/Arte_UFO_1_fr.htm

2/ CLOSE ENCOUNTER CLASSIFICATION

https://0knight.wordpress.com/2018/01/15/secretscreatures-and-visionaries/

3 / CHRONOLOGY OF MAIN EVENT OF UFO SIGHTINGS

https://ascensionavatar.wordpress.com/2019/06/28/lisa-renee-military-abduction/

http://ekladata.com/F_3f0N7vRAuHnb19iZGqVUAcKws/Dehlinger-Emmanuel_Ovnis_L-Armee-Demasquee.pdf

https://mysteriousworld.fandom.com/wiki/List_of_major_UFO_sightings#cite_note-9

https://www.smliv.com/stories/can-modern-science-help-solve-the-ancient-mystery/

https://fr.wikipedia.org/wiki/Chasseurs_fant%C3%B4mes

http://forgetomori.com/2010/ufos/the-varginha-incident-case-closed/#more-1676

https://fr.wikipedia.org/wiki/Cas_de_Carson_Sink#/media/Fichier:Carson(reconstitution).png

http://ovni-extraterrestre.com/l-observation-d-ovni-de-los-angeles-1942

https://worldofwarplanes.eu/fr/news/history-los-angeles-raid/

https://synchronicite.blog4ever.com/les-lumieres-de-brown-mountain

https://commons.wikimedia.org/wiki/File:LinkSeltsame_Gestalt_so_in_disem_MDLXVI_Jar.jpg

https://fr.wikipedia.org/wiki/Fichier:Nuremberg1561.jpg

https://www.bibliotecapleyades.net/ufo_aleman/esp_ufoaleman_10.htm

https://www.artstation.com/artwork/oOykKW

https://www.lamentiraestaahifuera.com/2010/10/23/el-ovni-de-canarias-del-79/

https://es.wikipedia.org/wiki/Incidente_del_misil_Poseidón_de_Canarias

https://www.rts.ch/info/sciences-tech/9198661-la-fusee-de-spacex-prend-des-allures-d-ovni-a-son-decollage-de-californie.html

https://www.history.com/news/lubbock-lights-ufo-sightings

http://www.ufo-alarm.com/index.php?sel=detail&item=89

https://fr.wikipedia.org/wiki/Incident_de_T%C3%A9h%C3%A9ran

https://en.wikipedia.org/wiki/Mariana_UFO_incident

https://www.angamen.com/disappearance-of-felix-moncla-and-the-unknown-radar-blip/

https://fr.wikipedia.org/wiki/Incident_de_l%27%C3%AEle_Maury

https://www.russianlover.site/en/Kapustin-and-the-51-area-of-%E2%80%8B%E2%80%8BRussia/

http://dossiers.secrets.free.fr/ovni/diamant.html#dia2

http://www.stuff.co.nz/marlborough-express/your-marlborough/9568965/The-Kaikoura-lights

https://www.tripadvisor.fr/Attraction_Review-g189835-d10696269-Reviews-UFO_monumentet-Angelholm_Skane_County.html
https://en.wikipedia.org/wiki/UFO-Memorial_%C3%84ngelholm

https://cryptozoo.pagesperso-orange.fr/dossiers/desmodus.htm

https://www.astonishinglegends.com/al-podcasts/2017/8/18/ep-79-the-kelly-hopkinsville-encounter-part-3

https://www.history.com/news/first-alien-abduction-account-barney-betty-hill

https://www.zersetzung.org/kt-citations?layout=blog&start=10

https://it.wikipedia.org/wiki/Caso_di_Trans-en-Provence

http://ovni-extraterrestres.over-blog.com/2014/06/l-affaire-de-trans-en-provence.html

https://en.wikipedia.org/wiki/Lonnie_Zamora_incident#Hoax_claims_and_rebuttals

https://fr.wikipedia.org/wiki/Disparition_de_Travis_Walton

https://www.amazon.com/Travis-True-Story-Walton/dp/B07HNJLDYY

https://it.wikipedia.org/wiki/Rapimento_alieno_di_Emilcin

http://www.abovetopsecret.com/forum/thread1051718/pg1

http://skepticversustheflyingsaucers.blogspot.com/2016/06/rencontre-rapprochee-ariel-school-ruwa.html

4/ "ALIEN" ABDUCTION & MILABS

https://www.bibliotecapleyades.net/vida_alien/alien_abductionabductees03.htm

https://ascensionglossary.com/index.php/MILABS

https://ordo-ab-chao.fr/ovni-abductions-abus-rituels-controle-mental/

5/ "ALIENS" IMPLANTS

https://www.theblackvault.com/casefiles/analysis-suspected-alien-implant-2/

https://www.theblackvault.com/casefiles/analysis-suspected-alien-implant-march-11-2006/

https://mercadonews.com/v5/index.php/Big-Brother/alien-implants-a-closer-look-into-one-aspect-of-alien-abduction.html

https://en.wikipedia.org/wiki/RFID

6/ ANIMAL MUTILATION

https://www.huffpost.com/entry/cattle-mutiliations_b_932711

7/ THE CROP CIRCLES

https://www.ovnis-armee.org/5_crop_circles.htm

https://www.savoirperdu.fr/articles/les-crop-circles/

8/ PLASMA & HOLOGRAPHIC TECHNOLOGY

https://www.ovnis-armee.org/11_plasma_technology.htm

https://www.wired.com/2007/05/plasma-laser-uf/

http://www.sweetliberty.org/issues/hoax/af.shtml

https://patentimages.storage.googleapis.com/ed/2c/03/851942fd7d803a/WO2005099386A2.pdf

http://www.historycommons.org/context.jsp?item=a1994projector#a1994projector

https://www.acalltoactions.com/single-post/2016/11/17/Patent-Documents-on-Project-Blue-Beam-Technology-The-False-Flag-Extra-Terrestrial-Invasion

https://apps.dtic.mil/dtic/tr/fulltext/u2/a392587.pdf

9/ STRANGE DEATH

http://ekladata.com/F_3f0N7vRAuHnb19iZGqVUAcKws/Dehlinger-Emmanuel_Ovnis_L-Armee-Demasquee.pdf

https://www.blueblurrylines.com/2018/05/capt-william-davidson-lt-frank-brown.html

https://eseforverdade.com.br/operacao-prato-ovnis-no-brasil/uyrange-holanda/

https://arquivoufo.com.br/2017/08/31/jim-e-coral-lorenzen-os-primeiros-pesquisadores-de-ovnis/

http://ovniparanormal.over-blog.com/2017/01/les-revues-ovnis-phenomenes-spatiaux-du-gepa-en-ligne-pdf.html

https://www.cohenufo.org/mcdnld_char.html

http://thebiggeststudy.blogspot.com/2010/04/fsr1959-human-side.html

https://factrepublic.com/facts/23563

10/ MISSILE DISABLED

https://www.dailymail.co.uk/news/article-7037549/Air-Force-deployed-20-missiles-fry-military-electronics-North-Korea-Iran.html

https://www.tomsguide.fr/boeing-un-missile-capable-de-desactiver-tout-appareil-electrique

11/ WAS FLYING SAUCER INVENTED BY NAZI?

https://cyprustar.wordpress.com/2017/12/13/operation-paperclip-secteur-aerospatial/

http://soldatsminiaturesdusergentgarcia.over-blog.com/pages/Ovnis-et-armes-secretes-nazies-6299827.html

12/ THE TR3B

https://www.matrixdisclosure.com/tr-3b-black-triangle

https://ovnis-videos.com/2016/06/03/enquete-sur-le-tr-3b

13/ OTHER ANTI GRAVITY SPACECRAFT

https://www.bibliotecapleyades.net/ciencia/ciencia_antigravity.htm

14/ THE USO

https://www.rt.com/news/russian-navy-ufo-records-say-aliens-love-oceans/

https://www.youtube.com/watch?v=rYMF5G9Npko

https://www.powervision.me/eu/product/powerray

https://phys.org/news/2013-12-navy-uav-submerged-submarine.html

15/ 70 YEARS OF PSYCHOLOGICAL WARFARE

https://www.winterwatch.net/2018/12/the-ufo-alien-narrative-is-a-psyop

http://www.policestateusa.com/2015/cia-admits-source-ufo-hysteria

https://en.wikipedia.org/wiki/Mirage_Men

https://www.hollywoodreporter.com/review/mirage-men-film-review-568349

https://fr.wikipedia.org/wiki/Steven_M._Greer

16/ THE SHADOW GOVERNMENT

https://www.bibliotecapleyades.net/sociopolitica/esp_sociopol_secretgov_2.htm

17/ ELECTRIC SENSIVITY TEST

http://ekladata.com/F_3f0N7vRAuHnb19iZGqVUAcKws/Dehlinger-Emmanuel_Ovnis_L-Armee-Demasquee.pdf

AUTHOR CONTACT

Mail:
cyprustar@gmail.com

Blog:
https://cyprustar.wordpress.com/

SUMMARY

INTRODUCTION	2
1/ UFO IN THE ART AND HISTORY?	3
2/ CLOSE ENCOUNTER CLASSIFICATION	17
3 / CHRONOLOGY OF MAIN EVENT OF UFO SIGHTINGS	19
4/ "ALIEN" ABDUCTION & MILABS	57
5/ "ALIENS" IMPLANTS	67
6/ ANIMAL MUTILATION	71
7/ THE CROP CIRCLES	75
8/ PLASMA & HOLOGRAPHIC TECHNOLOGY	83
9/ STRANGE DEATH	100
10/ MISSILE DISABLED	112
11/ WAS FLYING SAUCER INVENTED BY NAZI?	114
12/ THE TR3B	157
13/ OTHER ANTI GRAVITY SPACECRAFT	161
14/ THE USO	167
15/ 70 YEARS OF PSYCHOLOGICAL WARFARE	171
16/ THE SHADOW GOVERNMENT	186
17/ ELECTRIC SENSIVITY TEST	201
18/ DIGITAL SOURCES	206
AUTHOR CONTACT	214

www.ingramcontent.com/pod-product-compliance
Lightning Source LLC
Chambersburg PA
CBHW030622220526
45463CB00004B/1383